T0262177

Functional Genomics

Functional Genomics

Edited by **Rosanna Mann**

New York

Published by Callisto Reference,
106 Park Avenue, Suite 200,
New York, NY 10016, USA
www.callistoreference.com

Functional Genomics
Edited by Rosanna Mann

© 2015 Callisto Reference

International Standard Book Number: 978-1-63239-347-0 (Hardback)

This book contains information obtained from authentic and highly regarded sources. Copyright for all individual chapters remain with the respective authors as indicated. A wide variety of references are listed. Permission and sources are indicated; for detailed attributions, please refer to the permissions page. Reasonable efforts have been made to publish reliable data and information, but the authors, editors and publisher cannot assume any responsibility for the validity of all materials or the consequences of their use.

The publisher's policy is to use permanent paper from mills that operate a sustainable forestry policy. Furthermore, the publisher ensures that the text paper and cover boards used have met acceptable environmental accreditation standards.

Trademark Notice: Registered trademark of products or corporate names are used only for explanation and identification without intent to infringe.

Printed in the United States of America.

Contents

Preface

The purpose of the book is to provide a glimpse into the dynamics and to present opinions and studies of some of the scientists engaged in the development of new ideas in the field from very different standpoints. This book will prove useful to students and researchers owing to its high content quality.

Functional genomics is a field of molecular biology that allows the exploration of genes, interactions and protein functions on a universal level. This book highlights important matters in the branch of functional genomics, ranging from evaluation of the genetic codes, to understanding of the role of varied genes and proteomic ramifications. This book gives a perspective on primary problems and latest progresses in science and technology in the field of genomics medicine. Due to the vastness of the subject of functional genomics and further important sub-divisions like its applications, it is not possible to cover it all in one single book. Therefore, this book provides a short preface to various topics, dealing mainly with their relation to functional genomics and stresses on its particular uses in bio-medicine, agro-food technologies and zootechniques.

At the end, I would like to appreciate all the efforts made by the authors in completing their chapters professionally. I express my deepest gratitude to all of them for contributing to this book by sharing their valuable works. A special thanks to my family and friends for their constant support in this journey.

Editor

RNAi Towards Functional Genomics Studies

Gabriela N. Tenea and Liliana Burlibasa

Additional information is available at the end of the chapter

1. Introduction

RNA interference is an evolutionarily conserved mechanism that uses short antisense RNAs that are generated by 'dicing' dsRNA precursors to target corresponding mRNAs for cleavage. Pioneering observations on RNAi were reported in plants, but later on RNAi-related events were described in almost all eukaryotic organisms, including protozoa, flies, nematodes, insects, parasites, and mouse and human cell lines [1, 2]. Called initial cosupression or PTGS (post-transcriptional gene silencing), RNAi was first discovered in transgenic petunia plants [3]. In order to increase the pigmentation the chalcone synthase (CHS) gene was over-expressed in petunia plants and instead of enhancing in the flower pigmentation an opposite effect was observed. Some of the flowers were completely lacked of pigmentation and others showed different degrees of pigmentation. It was shown that even though an extra copy of the transgene was present, the CHS mRNA levels were strongly reduced in the white sectors. It was suggested that interaction between transgenes and native transcripts triggers a mechanisms that leads to the destruction of both transcripts or to the obstruction of the translation process and to gene silencing. This phenomenon was called co-suppression because the extra copies of CHS transgene determined reduction of its own expression but also the endogenous gene expression.

Later on, other study in the field of virus resistance was being exploited in order to produce virus resistance plants. Using different viral systems it has been shown that the expression of viral genes in the target plant genome was not associated with resistance to that particular virus [4-6]. The virus resistance in the recovered plants correlated with reduction of transgene mRNA in the cytoplasm, these phenomena was also called co-suppression. The finding provided supporting evidence of plant natural response to viral infection that the recovered parts of this plants response to virus would not only be resistant to initially inoculated virus but also cross-protect the plants against other viruses carrying homologous sequences [7]. This phenomenon was later called VIGS (virus-induced gene silencing). Further work found that the transcripts produced from both loci have been degraded in the cytoplasm. In this case, activation of PTGS was taught to be due to the production of aberrant dsRNA by the transgene, which results in the silencing of the mRNA [8].

In fungi *Neurospora crassa*, it was shown that an overexpressed transgene could induce gene silencing at the post-transcriptional level a phenomenon called "quelling"[9]. Few years later, the efficiency of injecting single-stranded anti-sense RNA as a method of gene silencing in the nematode *Caenorhabditis elegans* by using its ability to hybridize with endogenous mRNA and inhibits translation was investigated [10]. The discovery that introduction of a dsRNA was more effective at inhibiting the target gene led to the conclusion that both original single stranded sense and antisense samples, may have been initially were contaminated with dsRNA. However, similarities between plant and nematode were recognized and the term RNAi was adopted in both systems [11]. Initial experiments on RNAi were successfully in plants and nematodes by introduction of a dsRNA into the cytoplasm. In mammalian cells, however, similar techniques, results in the initiation of the interferon response and cell death before processing could occur. This happen till the group of Elbashir and co-workers [12] reported as alternative method by introduction of a siRNAs under 30 base pair length in the mammalian cells and the interferon response was avoided and activated the RISC (RNA interfering silencing complex) complex and the mRNA destruction.

The importance of the discovery of the RNAi by Fire and Mello was acknowledged in 2006 with the Nobel Prize in Physiology and Medicine. Shortly after this discovery, dsRNAs were found to induce similar gene silencing in a variety of other organisms: in the fruit fly *Drosophila* [13], zebrafish *Danio rerio* [14], *Hydra magnipapillata* (cnidarian) [15], and in some plant species [16-17]. Many experiments have shown that an intermediate in the RNAi process, called short-interfering RNAs (siRNA), might be effective in degrading mRNA in mammalian cells [18-19]. Nonetheless, it was still not believed that RNAi could work in humans, because long dsRNAs, larger than 30 base pairs in length, induce a cellular response (e.g. interferon response). The first evidence that RNAi functions in humans came from experiments performed by two groups of researchers. Kreutzer and Limmer (2000) [20] demonstrated that short fragments of dsRNA might mediate the RNAi response triggered by the long dsRNAs as observed by Fire and Mello. Despite the fact that these findings were not published, a key patent around this discovery and a company focused on the development and commercialization of RNAi therapeutics was established (www.huntington-assoc.com). In the same time, Elbashir and co-workers [21] found that synthetic short dsRNA molecules result in a potent RNAi gene silencing in mammalian cells without inducing interferon response.

The knowledge accumulated from RNAi studies opened an enormous potential for the use as a tool in functional genomics studies in both plant and animal systems. In recent years, numerous strategies have been developed for targeted gene silencing and a combination of approaches enhanced the manipulation of gene silencing for functional genomics studies.

2. The molecular mechanism of RNA interference in eukaryote system

RNA silencing mechanism was first recognized as antiviral mechanism that protects organisms from RNA viruses, or prevents random integration of transposable elements [22-

25, 26]. In the last few years, important insights have been gained in elucidating the molecular mechanism of RNAi by identification and characterization of the central players of the core RNAi pathway. Extensive genetic and biochemical studies in various species have yielded a common model of RNA silencing in which trigger dsRNA. Using genetic screening analyses performed in several organisms, such as the fungus *Neurospora crassa*, the alga *Chlamydomonas reinhardtii*, the nematode *Caenorhabditis elegans*, and the plant *Arabidopsis thaliana* several host-encoded proteins involved in gene silencing as well as the essential enzymes or factors common in this process has been identified [27-28]. The molecular mechanism of RNA silencing involves several steps and a key step of silencing is the processing of dsRNAs precursors into short RNA duplexes [29-30].

2.1. The core RNAi mechanism

2.1.1. Processing the dsRNA precursors

In the initiator step, the enzyme Dicer (RNAse III-like enzyme) chops dsRNA into small pieces called short interfering RNA (siRNA), which are around 21-24 nucleotide in length [12, 21, 26, 31-33]. Dicer or Drosha, proteins known for their catalytic RNAseIII and dsRNA-binding domains, catalyzes the maturation of small RNAs. miRNAs are transcribed as long primary transcripts, which are processed by Drosha in the nucleus. Nuclear transport occurs through nuclear pore complexes, which are large proteinaceous channels deposited in the nuclear membrane. The miRNA precursor in further transported to the cytoplasm by means of the nuclear export receptor, exportin-5. Following their export from the nucleus, pre-miRNAs are subsequently processed by the cytoplasmic Dicer that yields RNAs duplexes of 21 nucleotides in length, with 5' phosphates and 2-nucleotide 3' overhangs. Numerous Dicer proteins have been identified in plants as well as animal system and each Dicer is preferentially processing dsRNAs, which comes from different sources [26].

2.1.2. RNA silencing effector complex assembly

siRNAs are loaded into the effector protein complex to form an RNA-induced gene silencing complex, called RISC-complex. Subsequently, the siRNA within RISC unzips, exposing anticodons and thus activating the RISC. Usually, effector complexes containing siRNAs are known as a RISC, while those containing miRNAs are known as miRNPs. For example, in *Arabidopsis thaliana* the rasiRNA-containing effector complexes are known as RITSs. All RISCs or miRNPs have a member of the Argonaute (Ago) family of proteins attached to them. RISCs and miRNPs differ in size and composition, based on the provenience organism. Further studies for identification of more specific and active siRNA duplexes for guidance of cleavage of mRNA, revealed that the sequence of the siRNA duplex had a significant impact on the ratio of sense and antisense siRNAs that were entering the RISC complex [26]. There have been different numbers of Ago proteins identified in different organisms. *Arabidopsis thaliana* has ten members, *D. melanogaster* has five members, and humans have eight members of the Ago protein family. Only a small number of these proteins have actually had their function characterized [26].

2.1.3. mRNA cleavage and repression of translation

After the formation of the RISC complex, the siRNAs in the RISC complex guide degradation that is sequence-specific, of the complementary or near complementary mRNAs [26]. The RISC complex cleaved the mRNA in the middle of its complementary region. The cleavage does not require the presence of ATP, however multiple cleavages are more efficient in the presence of ATP. RISC and miRNP complexes work by catalyzing hydrolysis of the phosphodiester linkage of the target RNA [26]. It is not fully understood the mechanism by which repression of translation guided by miRNA (micro-RNA) as well as the mechanism by which mRNA cleavage are working. The first evidence that this occurs was described in *C. elegans* mutants, where specifically targeted miRNAs reduced synthesis of proteins without affecting the levels of mRNA. It has been suggested that miRNAs affect translation termination or elongation rather than actual initiation of the process. In addition, it has been shown that miRNAs can act as siRNAs and vice versa. Further investigations suggested that mRNA degradation and translational regulation guided by miRNAs could be used as simultaneous mechanisms for natural regulation [26]. Notably, siRNAs can also move from cell to cell and systematically spread and deliver the silencing signal to the entire organism [34].

2.2. RNA silencing pathways in mammals

The 21-nucleotide miRNAs derive from dsRNA-like hairpin regions of 70 nucleotides within primary transcript [35]. Firstly, cleavage of the pri-miRNA in the nucleus by the RNAse III enzyme Drosha releases the stem-loop (or hairpin), and this precursor (pre-miRNA) is subsequently exported to the cytoplasm. The end of the stem of the pre-miRNA has the same characteristic 5′and 3′termini as siRNAs. In the cytoplasm a Dicer enzyme makes a pair of cuts that liberates a 21–nucleotide RNA duplex. Similar to siRNA duplexes, the strand whose 5′end is less stably paired will be used as guide/miRNA strand [36]. miRNA and RNAi pathways share the same core machinery, but in various animal species exist different specialization. MicroRNAs and non-coding RNAs are a major breakthrough in epigenetic of the last years, and have been found to contribute to almost all biological pathways, including gametogenesis, early development and cell signaling. While, this RNA gene silencing pathway is used by both siRNAs and miRNAs, there exist some important differences.

Comparison of *Drosophila, C. elegans* and humans has revealed that homologous Drosha enzymes catalyze the first processing step of their miRNA pathway [35]. Nonetheless, these three species show more variation in respect to the functional roles of Dicer enzymes. In *Drosophila,* two distinct enzymes are responsible for pre-miRNA cleavage and siRNA production [26]. In *C. elegans* and humans only one Dicer enzyme is present, having both cleavage functions [26].

In humans and other vertebrates, the main RNA silencing pathway seems to involved miRNA because of the existence of an immune response against long dsRNA, suggesting that the processing pathway for this type of RNAi trigger would be less important [37]. In humans, four Ago proteins (ago 1-4) have been identified [38]. A study of the assembly of human RISC has revealed that miRNA processing and Ago2-mediated target-RNA cleavage are functionally coupled [39]. The demonstration of physical and functional coupling of pre-miRNA processing and target-RNA cleavage provides an explanation for earlier observations that 27 nucleotide dsRNA and short hairpin RNA (shRNA) are considerably more potent triggers of RNAi than duplex siRNA [40-41].

Animal miRNAs may act combinatorial, several miRNAs could binding a single transcript [40]. Also, experiments performed by Doench and his co-workers (2003) [42] suggested that multiple miRNAs could act cooperatively, reducing mRNA translation by more than the sum of their individual effect. The high number of putative target genes indicates that miRNAs function in a broad range of biological processes. Until now, the function of only a few miRNA have been analyzed in animals, but these studies have already revealed important roles of miRNAs in control of cell division, differentiation, apoptosis [43] in several developmental processes such as morphogenesis, neurogenesis and developmental timing [44, 45]. Nonetheless, antisense RNA has been implicated in imprinting and X inactivation.

Studies of miRNA processing have also provided information that could improve scientists ability to design more efficient RNAi inducing RNA molecules for experimental and therapeutic applications [46]. Krutzfeldt and co-workers [47] identified a novel class of chemically engineered oligonucleotides named "antagomirs" which silenced miRNA in murine model. Antagomirs are cholesterol-conjugated single-stranded RNA molecules 21-23nt in length and complementary to the mature target miRNA [48]. These oligonucleotides can be designed to specifically bind to the miRNA-RISC complex and they inhibit its function. Antagomirs down-regulated the proteins translation by the silencing miRNA [48]. It has been shown that these molecules are very stable *in vivo* and after one intravenous injection only, and can silence target miRNA in the liver, lung, intestine, heart, skin and bone marrow for more than a week.

2.3. RNA silencing pathways in plants

RNA gene silencing was discovered in plants as a mechanism whereby invading nucleic acids, such as transgenes and viruses, are silenced through the action of small (20–26nt) homologous RNA molecules [3]. Crucial to understanding the gene silencing mechanism is to know how to trigger it from the theoretical perspective of understanding a remarkable biological response to the practical use of silencing as an experimental tool.

This process is initially triggered by dsRNA, which can be introduced experimentally or arise from endogenous transposons, replicating RNA viruses, or the transcription of the transgenes as shown in Figure 1. In brief, double-stranded RNAs generated through aberrant gene expression from a foreign gene, virus infection or tandem repeats sequences due to insertion of transposons / retrotransposon are digested into 21-25 nucleotide long siRNAs by Dicer. This siRNA functioned as a template for the targeted degradation of mRNA in RISC and acts also as the primer for RdRP to amplify the secondary dsRNA [49]. As mentioned in the mammals system numerous components of RNAi machinery were identified and characterized in plants. For example, Argonaute proteins played an important role in RNA silencing in plants because they are components of the silencing effector complexes that bind to siRNAs and miRNAs. Dicer proteins are required for miRNAs biogenesis. Unlike the animal system, miRNAs in plants are more paired to their target RNA and use RNA cleavage rather than translational suppression as the primary silencing mechanism [35].

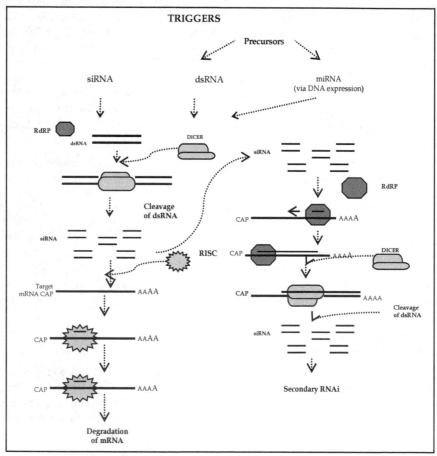

Figure 1. RNA mediated gene -silencing pathway. Double stranded RNA is digested in siRNAs by Dicer; this siRNA function as a template for the target degradation of mRNA in RISC. siRNA acts as the primer for RdRp to amplify the secondary dsRNA. RISC (RNA inducing silencing complex); Dicer (RNaseIII –like RNase); RdRP (RNA-dependent RNA polymerase)

In plants RNAi process engages the participation of numerous pathways [32]. Diverse biological roles of these pathways have been established including the mechanism of viral defense, regulation of gene expression and the condensation of chromatin into heterochromatin. The first pathway of RNA silencing, called cytoplasmic siRNA silencing, is a mechanism by which the dsRNA could be a replication intermediate or a secondary-structure feature of single-stranded viral RNA and maybe important for virus-infected plant cells. The source of dsRNAs includes replication intermediates of plant RNA viruses, transgenic inverted repeats, and products of RNA-dependent RNA polymerases (RdRPs). The dsRNA may be form by annealing of overlapping complementary transcripts [30].

Silencing of endogenous messenger RNAs by miRNAs is a second pathway of silencing in plants. These miRNAs negatively regulate gene expression by base pairing to specific mRNAs, resulting in either RNA cleavage or arrest of protein translation. Like siRNAs, the miRNAs are short 21-24-nucleotide RNAs derived by Dicer cleavage of a precursor [32]. miRNAs downregulate gene expression through base-pairing to target mRNAs, leading to either the degradation of mRNAs or the inhibition of translation or both. In plants, the prototype miRNAs were identified as a subset of the short RNA population, and are derived from an inverted repeat precursor RNA with partially double-stranded regions, and they target a complementary single-stranded mRNA.

The third pathway of RNA silencing in plants is associated with DNA methylation and suppression of transcription. This type of silencing was evidenced in plants by the discovery that the transgene and viral RNAs guide DNA methylation to specific nucleotide sequences. More recently, these findings have been extended by the observations that siRNA-directed DNA methylation in plants is linked to histone modification [30]. An important role of RNA silencing at the chromatin level is probably protecting the genome against damage caused by transposons.

2.4. Types of small RNAs in eukaryote

2.4.1. siRNAs

Small interfering RNAs (siRNAs) are 22nt fragments, which bind to the complementary portion of their target mRNA and tag it for degradation. siRNA have a role in conferring viral resistance and secures genome stability by preventing transposon hopping. RNAi mechanisms, in which siRNAs are involved keep chromatin condensed and suppress transcription, repress protein synthesis and regulate the development of organisms.

2.4.2. miRNAs

miRNAs are integral components of the genetic program that account for approximately 5% of the predicted genes in plants, worms and vertebrates. Loci encoding these miRNA are localised in the introns of protein-coding genes or in the noncoding region of the genome [50]. miRNAs help regulate gene expression, particularly during development [51]. The phenomenon of RNA interference, broadly defined, includes the endogenously induced gene silencing effects of miRNA as well as silencing triggered by foreign dsRNA. Mature miRNAs are structurally similar to siRNAs produced from exogenous dsRNA, but before reaching maturity, miRNAs must first undergo extensive post-transcriptional modification.

miRNA is express from a longer RNA coding locus as a primary transcript called pri-miRNA which is processed in the nucleus, to a 70-nucleotide hairpin structure known as pre-miRNA. The small miRNAs are processed from larger hairpin precursors by an RNAi-like machinery.

The first miRNA, lin-4, was discovered in C. *elegans* five years prior to the demonstration of dsRNA as an inducer or RNAi [52]. Short non-coding transcript from lin-4 represses the expression of the nuclear protein encoding gene lin-14 as part of the control of developmental timing. The existence of partial complementarity between the small lin-4 RNA and several elements in the 3' untranslated region (UTR) of the lin-14 mRNA suggested a mechanism of translational inhibition via an antisense RNA-RNA interaction. In this context, miRNAs were shown to compose a large class of ribo-regulators [36, 53-54]. In the same time, was demonstrated that Dicer converts pre-miRNA into mature miRNAs of approximately the same length as single-stranded siRNAs, establishing a formal connection between miRNAs and siRNAs [55-56]. Other studies have revealed the complete pathway of miRNA processing in animals, which is based on two steps catalysed by the RNase III enzymes Drosha and Dicer. The mature miRNAs of animals generally regulate their target genes by translational repression, but some cases of target mRNA cleavage have also been reported [35, 57]. This is in contrast with the situation in plants, in which target mRNA cleavage appears to be the main mechanism [58].

With the discovery of the first miRNA lin-4, interest in the role of miRNA in the regulation of fundamental biological processes has rapidly emerged. Now, more than 18226 entries representing hairpin precursor miRNAs, expressing 21643 mature miRNA products, in 168 species are tabulated in the miRNA registry (http://microrna.sanger.ac.uk). Among them, more than 300 miRNA have been discovered in humans [46]. In mammals, about one-half of the know miRNA are located within the transcription units of other genes and share a single primary transcript [59-60]. These miRNAs generally reside in the introns or in exon sequences that are not protein coding. The expression pattern of the miRNA varied. While some C. *elegans* and *Drosophila* miRNAs were expressed in all cells and at all developmental stages, other had a more restricted spatial and temporal expression pattern. This suggested that such miRNAs might be involved in post-transcriptional regulation of developmental genes [18].

In plants, as in animal systems, miRNAs, are generated as single-stranded 20-24-nucleotide species, by several proteins such Dicer and Argonaute (Ago). Ago proteins are components of the silencing effector complexes that bind the siRNAs and miRNAs. miRNAs act in *trans* on cellular target transcripts to induce their degradation via cleavage, or to attenuate protein production. Based on a computational genome analyses in *Arabidopsis*, it has been estimated that there are about 100 miRNA loci and some of them are conserved between *Arabidopsis* and *Lotus, Medicago* or *Populus* but not founded in rice [30]. Currently, there are numerous known plant miRNAs, and, in several cases, the target mRNA has been experimentally validated by expression of a miRNA-resistant target gene with silent mutations in the putative miRNA complementary region [30]. In *Arabidopsis* many miRNAs have been identified and correspond to the mRNAs for transcription factors and other proteins involved in gene regulation [61-63]. For example, *miR159* and its putative target transcription factor MYB33vmRNA, has been regulated by the hormone gibberellic acid [27]. Gibberellic acid stimulus could lend to an increase in MYB33 mRNA that would initiate flowering, and, directly or indirectly, to an increase in *miR159*. A similar mechanism has been identified for miR177 that target a transcription factor in GRAS mRNA [62].

2.4.3. Other molecules involved in RNAi processing

In addition to endogenous miRNAs and exogenous siRNAs, several other classes of siRNAs such as: *trans*-acting siRNAs (tasiRNAs), repeat-associated siRNAs (rasiRNAs), small-scan (scn)RNAs and Piwi-interacting (pi)RNAs have been identified. tasiRNAs are small (~21nt) RNAs that have been reported in plants, and they are encoded in intergenic regions that correspond to both the sense and antisense strands [64-65]. In *Arabidopsis thaliana*, tasiRNAs require components of the miRNA machinery and cleave their target mRNAs in *trans* [64-65]. These siRNAs represses the gene expression though post-transcriptional gene silencing in plants and it is transcribed from the genome to form a polyadenylated, double-stranded precursor. rasiRNAs that match sense and antisense sequences could be involved in transcriptional gene silencing in *Schizosaccharomyces pombe* and *A. thaliana* [66-68]. scnRNAs are ~28-nt RNAs that have been found in *Tetrahymena thermophila* and that might be involved in scanning DNA sequences in order to induce genome rearrangement [69]. piRNAs are different from miRNAs and are possibly important in mammalian gametogenesis [70]. They are small (~26–31-nt) RNAs that bind to MILI and MIWI proteins, a subgroup of Argonaute proteins that belong to the Piwi family and that are essential for spermatogenesis in mice.

2.5. Factors and proteins involved silencing

2.5.1. Dicers

Members of Dicer family which showed specificity for cleavage of dcRNAs, played central role in the processing of the dsRNAs precursors: miRNA and siRNA. Processing of dsRNAs by Dicer yields RNA duplexes of 21 nucleotides with 3′ overhangs of 2 to 3 nucleotides and 5′-phosphate and 3′-hydroxyl termini [12]. Dicer namely DCR (in *Drosophila*) / DCL (*Arabidopsis*), has four distinct domains: an amino terminal helicase domain, dual RNase III motifs, a dsRNA binding domain, and a PAZ domain (a 110-amino-acid domain present in proteins like Piwi, Argo, and Zwille/Pinhead), which it shares with the RDE1/QDE2/Argonaute family of proteins that has been genetically linked to RNAi by independent studies [71-72]. Cleavage by Dicer is thought to be catalyzed by its tandem RNase III domains. Some DCR proteins, including the one from *D. melanogaster*, contain an ATP-binding motif along with the DEAD box RNA helicase domain. In *Arabidopsis thaliana*, four Dicer-like proteins (DCL1, DCL2, DCL3 and DCL4) have been identified and are involved in the processing of several dsRNAs coming from different sources [73]. For example, DCL2 is required for production of siRNA from plant viruses while DCL3 is involved in production of rasiRNA [73]. On the other hand, in *C. elegans* and mammals one single Dicer gene, DCR-1 has been identified.

2.5.2. RISC complex

RICS complex, the effector of RNAi silencing is a multi-protein complex of which several components were identified. One of the proteins identified in almost all organisms is AGO protein that is essential for mRNA silencing activity [74]. In plants 10 AGO member proteins have been identified. For example, AGO1 mutant plants have been found to develop distinctive developmental defects. miRNAs are accumulated in these mutants but the cleavage of target mRNA not longer occur [75]. AGO4 has role in the production of long siRNAs of 24bp and it was early reported that AGO4 is involved in long siRNA mediated chromatin modifications (histone methylation and non-CpG DNA methylation) [76]. In addition to AGO family members, several other proteins associated with RISC complex have been identified in vertebrate and invertebrate models. For example, the *Drosophila* homologue of the fragile X mental retardation protein (FMRP); R2D2, found in *Drosophila* and thought to facilitate the passage of the Dicer substrate to the RISC; members of the mammalian Gemin family, some of which are thought to have helicase activity [44].

2.5.3. RNA-directed RNA polymerase (RdRP)

In both plants and *C. elegans*, RNAi/PTGS requires proteins similar in sequence to a tomato RNA-directed RNA polymerase [77]. In *Arabidopsis*, RdRP homologue SDE1/SGS2 is required for transgene silencing, but not for virally induced gene silencing [78]. This may suggest that SDE1/SGS2 act as an RdRP, since viral replicases could substitute for this function in VIGS. In *Neurospora*, RdRP homologue QDE-1 is required for efficient quelling [79]. EGO-1, one of the *C. elegans* RdRP, is essential for RNAi in the germline of the worm [80], and another RdRP homologue, RRF-1/RDE-9, is required for silencing in the soma. All RdRP proteins could be involved in amplifying the RNAi signal. However, only the tomato and *Neurospora* enzymes have been demonstrated to posses RNA polymerase activity, and biochemical studies are required to establish definitively the role of these proteins in RNAi [81].

2.5.4. Putative helicase

Other proteins, helicases have been identified in plants (e.g *sde3* in *Arabidopsis*) [78]. Although possible roles in RNAi for some of these proteins were proposed, e.g. MUT6 might involved in the degradation of misprocessed aberrant RNAs [81], their functions are mostly unknown and further biochemical experiments are needed to reveal their exact roles in RNAi. The quelling-defective mutant in *Neurospora*, *qde3*, was cloned and the sequence encodes a 1,955-amino acid protein. This protein shows homology with the family of RecQ DNA helicases, which includes the human proteins for Bloom syndrome and Werner syndrome.

3. Applications of RNAi in plant systems

RNAi has been used as new tool to reduce the expression of a particular gene in mammalian and plant cell systems, to analyze the effect that gene has on cellular function, and also it has the potential to be exploited therapeutically and clinical trials. However, by using RNAi, scientists can quickly and easily reduce the expression of a particular gene in mammalian and plant cell systems, often by 90% or greater, to analyze the effect that gene has on cellular function [18, 49, 82].

3.1. Development of efficient RNAi vector cassettes

In plants, many efforts were concentrated on the improvement of the nutritional content using the classical breeding approaches such as selection of the natural or induced genetic variations, or by genetic engineering of transgenic plants [83]. Genetic engineering technologies have advantages over classical breeding not only because they increase the scope of genes and the types of mutation that can be manipulated, but also because they have the ability to control the spatial and temporal expression patterns of the genes of interest [84]. The delivery of siRNAs in plants has been always achieved by expressing hairpin RNAs that fold back to create a double-strand region that will be recognized by the Dicer-like enzyme. Figure 2 depicted an typically RNAi construct in plants with the promoter region, the inverted repeats of the target gene with the appropriate orientation, the spacer region which separate the two inverted repeats sequences and the terminator region.

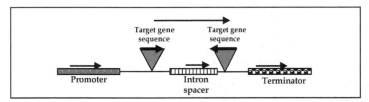

Figure 2. A schematic representation of a RNAi vector cassette with the promoter/ terminator region, target inverted repeats, intron spacer region; the arrows represent the direction of transcription

The double stranded RNA generated through aberrant gene expression from a foreign gene, viral infection, tandem repeat sequences formed due to insertion of a transposon/retrotransposon, are recognized by Dicer and digested in small interfering RNAs, which functioned as template for the targeted degradation of mRNA in RISC. This siRNA functioned also as primer for RdRP to amplify secondary dsRNA. On the other hand, in plants, RNAi is both systemic and heritable and siRNAs move between cells through channels in cell walls, thus enabling communication and transport throughout the plant. In addition, methylation of promoters targeted by RNAi confers heritability, as the new methylation pattern is copied in each new generation of the cell [85].

Genetic transformation via *Agrobacterium* or by particle bombardment or by infecting plants with viruses that can express the dsRNAs, or the infiltration of *Agrobacterium* harboring the hairpin cassette for transient gene silencing are the common methods for inducing gene silencing in plant system. The transgene expression should be evaluated as soon as possible for each transgenic event, and over multiple generations to insure that each line is stable-silencing its target. Many transgenic events should be generated and analyzed and the lines with active transgenes that are effectively inducing silencing cal be selected and maintained.

Currently, several vectors used for RNAi silencing that make use of *Agrobacterium* mediated delivery or artificially introduced dsRNA and/or VIGS into plants has been reported. For example, in 2007, an *Arabidopsis* genomic RNAi knock-out line analysis consortium was lunched out (AGRIKOLA) which is using the PCR products to generate gene-specific RNAi constructs for each *Arabidopsis* gene used in large scale gene silencing studies [86-87]; other consortium called CATMA (Complete *Arabidopsis* Transcriptome MicroArray), is generating gene sequence tags (GSTs) representing each *Arabidopsis* gene, designed so that they will hybridize on *Arabidopsis* cDNA microarrays in a gene-specific manner; the *Medicago truncatula* RNAi database (https://mtrnai.msi.umn.edu/) is a NSF-funded project planning to silence 1500 genes involved in symbiosis in this model legume; amiRNAi Central (http://www.agrikola.org) a NSF project funded to provide a comprehensive resource for knockdown of *Arabidopsis* genes.

Moreover, a set of binary vectors, called ChromDB's RNAi vectors, were designed for producing dominant negative RNAi mutants using a target sequence cloning strategy that is based on the inclusion of two restriction enzyme cleavage sites in each of two primers used to amplify gene-specific fragments from cDNA. This design minimizes the number of PCR primers and results in the placement of unique restriction enzyme recognition sites to allow for flexibility in future manipulations of the plasmid, *e.g.*, moving the inverted repeat target sequence to a different vector (Chrom database). These vectors are based on pCAMBIA binary vectors, a set of plasmids developed by the Center for Application of Molecular Biology to International Agriculture (CAMBIA).

The pHELLSGATE, high-throughput gene silencing vector and a high throughput tobacco rattle virus (TRV) based VIGS vector are binary vectors developed by Invitrogen are used for expression of GUS and GFP proteins. These vectors are base on Gateway recombination-based technology, which replaced the conventional cloning strategy. It is based on the phage lambda system of recombination. It enables segments of DNA to be transferred between different vectors while orientation and reading frame are maintained. It can also be used for transfer of PCR products. It saves valuable time, because once the DNA has been cloned into a Gateway vector, it can be used as many genome function analysis systems as is required. In this way, the use of vectors in the process of plant functional genomics has been made much easier, while the process has also been made faster. This allows for higher throughput analysis to occur [88].

3.2. RNAi and functional genomic studies

An important application of RNAi for functional genomics studies is to generate lines that are deficient for the activity of a subset of genes and then test the knockdown lines for a specific phenotype. The assessing of a specific phenotype requires the presence of a specific allele of marker genes and several generation of crosses are necessary for selecting a specific mutant allele for specific genotype. RNAi technology for functional genomics has advantage that a specific gene can be silenced if the target sequence is better chosen. However, since RNAi is a homology-dependent process a carefully selection of a unique or conserved region of the target gene ensures that a specific member of a multiples gene family can be silenced. For example, RNAi can down-regulate specific target sequences when 3'UTR region is used as a trigger sequence [89-90]. It has been also shown that RNAi facilitates the generation of dominant loss-of-function mutation in polyploidy plants, even with short dsRNAs of a 37 nucleotide long [91].

Nowadays, numerous projects are lunched to produce siRNAs that will silence essential genes in insects, nematodes and pathogens using an approach called hdRNAi (host-delivered RNAi) [92-94] based on the partial sequences similarities between plant and animal genes. There is also a limitation of this approach because, some unexpected genes can be silenced with consequences on the organisms itself but also environment.

RNAi has been also used for the improvement of nutritional value of some important crops. For example, to decrease the levels of natural toxins in food plants a stable, heritable and distinct siRNA against the toxin could be used. Cottonseeds are rich in dietary proteins but unpalatable by humans as they contain a natural toxic terpenoid item, called gossypol. RNAi mechanism has been used to minimize the levels of delta-cadinene synthase, an enzyme crucial for the production of gossypol [95].

RNAi technology has been also applied to barley for developing varieties resistant to BYDV (barley yellow dwarf virus) [96]. In rice, RNAi has been used to reduce the level of glutein and produce rice varieties with low-glutein content [97]. Soybeans can be engineered to produce oil with low levels of polyunsaturated fatty acids through a reduction of FAD2, a fatty acyl $\Delta 12$ desaturase. This enzyme converts the monounsaturated fatty acid oleic acid ($18:1\Delta 9$) to linoleic acid ($18:2\Delta 9$, $\Delta 12$), which can be subsequently desaturated to α-linolenic acid ($18:3\Delta 9$, $\Delta 12$, $\Delta 15$) by FAD3 [98]. The reduced polyunsaturated fatty acid levels from >65% of the total oil content in normal soybean oil to less than 5% was observed [99]. In an attempt to specifically target FAD2-1, and not related family members, the soybean FAD2-1A intron was tested as an RNAi trigger, resulting in a reduction in polyunsaturated fatty acids in the seeds to about 20% [100]. This result was surprising, given that intron sequences are removed from precursor mRNAs (pre-mRNAs) by splicing in the nucleus and spatially separated from the cytoplasm where mature mRNAs are presumed to be targeted by the PTGS machinery [101].

There are also some limitations of using RNAi in functional genomics studies. Unlike the insertional mutagenesis, for the use of RNAi the exact sequence of the target gene is required. Secondly, the methods to delivery RNAi are very important, some species are easily transformable and some not. Nonetheless, further improvement of the delivery methods and vectors that can be used safely and reliably are needed. There have also been some reports that revealed the difficulty to detect mutants with subtle changes in gene expression. However, in plants, numerous marker genes are being developed that will indicate if a change in gene expression occurs [102].

3.3. RNAi and viral infections

RNA silencing in plants prevents viral accumulation and accordingly, viruses have evolved several strategies to counteract the defense mechanism. A viral protein, HC-Pro (helper component proteinase) was shown to mediate one class of viral synergism disease [103] and expression of this protein in transgenic plants allows the accumulation of heterologous viruses beyond the normal level suggesting that HC-Pro blocked the target plant defence mechanism [104]. There are several methods known to identify viral suppressor proteins, such as transient expression assay, the reversal of silencing assay and stable expression assay.

A well known used method to study the transient expression is co-infiltration method using *Agrobacterium* strains, one strain used for inducing of RNA silencing of a reporter gene such as GFP (green fluorescent protein) and one strain that will express the candidate suppressor gene. Both strains will be infiltrated in a plant tissues such tobacco leaves, which are suited for production of a higher amount of protein in response to agro-infiltration. However, if the local silencing is triggered three days after infiltration the effect can be evaluated under UV light. If the candidate suppressor expressed from the co-infiltrated *Agrobacterium* interferes with RNA silencing, the tissue will remain bright green and in case not, the tissue will turn red [105]. In the case of reversal approach the candidates that may suppress silencing are identified. Several studies have shown that the viral suppressor proteins play an important role in this defense mechanism [106]. For two suppressor proteins, p21 encoded by beet western yellow virus [107] and p19 encoded by the tomato bushy stunt virus (TBSV) group [108] the molecular mechanism was identified.

Stable expression assay approach, a stable RNAi line expressing a suppressor candidate is crossed with several lines silenced for a repressor gene [109-110]. This method is also advantageous because provide information about the molecular mechanisms of the suppression and is also suited to investigate the role of suppression in systemic silencing using grafting [111].

However, the findings that certain viral proteins suppress RNA silencing open a new tool for biotechnologies applications. With silencing under control, many transgenic plants can be generated to produce desired plant traits or very higher level of expression to use the plant as a factor for producing pharmaceutical compounds, vaccines and other gene-products.

4. Application of RNAi in animal systems

4.1. RNAi and medicine

The ability to trigger RNAi in mammals was first demonstrated by microinjection of long dsRNA into mouse oocytes and one-cell stage embryos [19]. In this case was demonstrated that the antiviral interferon response to long dsRNAs is not yet functional in early mouse embryos. It was discovered rather quickly that chemically synthesized siRNAs could trigger sequence-specific silencing in cultured mammalian cells without inducing the interferon response [21]. Starting from this important breakthrough, RNAi has emerged as a powerful experimental tool for analyzing mammalian systems.

The ability of RNAi to determine ablation of gene expression has open up the possibility of using collections of siRNAs to analyze the role of hundreds or thousands of different genes whose expression is know to be up-regulated in a disease, given an appropriate tissue culture model of that disease. The libraries of RNAi reagents can be used in one of two ways. One is in a high throughput manner, in which each gene in the genome is knocked down one at a time. The other approach is to use large pools of RNA interference viral vectors and apply a selective pressure that only cells with the desired change in behavior can survive [112]. Rapid progress in the application of RNAi to mammalian cells, including neurons, muscle cells, offers new approaches to drug target identification. Advances in targeted delivery of RNAi-inducing molecules has raised the possibility of using RNAi directly as a therapy for a variety of human genetic disorders.

4.2. RNAi and therapy

Considering the gene-specific features of RNAi, it is conceivable that this method it will be very useful for therapeutic applications. Direct transfection of siRNAs into cells, creating an expression construct in which a promoter drives the production of both the sense and antisense siRNAs which then hybridize in the cell to produce the double stranded siRNA and using viral vectors to infect cells with an expression construct are the methods used nowadays for RNAi-based therapy.

Nonetheless, this hypothesis is based on the assumption that the effect of exogenous siRNA applications will remain gene specific and do not show nonspecific side effects relating to mismatches off-target hybridization or protein binding to nucleic acids. For example, several research groups have explored the use of RNAi to limit infection by viruses in cultured cells. There is a huge potential for using RNAi for the treatment of viral diseases such as those caused by the human immunodeficiency virus (HIV) and the hepatitis C virus.

RNAi strategy includes multiple targets to neutralize HIV. For example, directed siRNAs against several regions of the HIV-1 genome, including the viral long terminal repeat (LTR) and the accessory genes, *vif* and *nef* [113-115]. Using Magi cells (CD4-positive HeLa cells) as a model system, they demonstrated a sequence specific reduction of >95% in viral infection after cotransfection of siRNAs with an HIV-1 proviral DNA. When the same assay was done in primary peripheral blood lymphocytes, which are natural targets for HIV-1, the frequency of infected cells was also substantially reduced. These could be targets that block entry into the cell and disrupts the virus reproduction cycle inside the cells. This technology will help researchers dissect the biology of HIV infection and design drugs based on this molecular information [116]. Researchers from Hope Cancer Center in Duarte have developed a DNA-based delivery system in which human cells are generated to produce siRNA against REV protein, which is important in causing AIDS [117].

The delivery of siRNA to HIV-infected T lymphocytes, monocytes and macrophages is a challenge. As synthetic siRNAs do not persist for long periods in cells, they would have to be delivered repeatedly for years to treat the infection. Systemic delivery of siRNAs to lymphocytes is not feasible owning to the huge number of these cells. Therefore, the preferred method is to isolate T cells from patients. In clinical trial T cells from HIV-infected individuals are transduced ex vivo with a lentiviral vector that encodes an anti-HIV antisense RNA, and then reinfused into patients [118]. In other study, it was reported that the GFP siRNA induced gene silencing of transient or stably expressed GFP mRNA was highly specific in the human embryonic kidney (HEK) 293 cell background [119]. Further study, in human non-cell lung carcinoma cell line H1299 have shown that specific siRNAs corresponding to *akt1*, *rb1*, and *plk1* could be used as highly specific tools for targeted gene knockdown and can be used in high-throughput approaches and drug target validation.

RNAi is also utilized as an antiviral therapy against diseases caused by herpes simplex virus type 2, hepatits A and hepatits B. Early RNAi studies noted that RNA silencing was prominent in the liver, which made this organ an attractive target for therapeutic approaches. Vaccine against HBV is used only for prevention and there is no vaccine for HCV. Mc Caffrey and his co-workers (2003) [120] demonstrated that a significant knockdown of the HBV core antigen in liver hepatocytes could be achieved by the siRNA, providing an important proof of principle for future antiviral applications of RNAi. They developed a transient model of HBV infection in which a plasmid containing approximately 1-3 copies of the HBV genome (pTHBV2) was introduced into the livers of mice by hydrodynamic transfection. This results in production of all four families of viral mRNAs, including the pregenomic RNA. The pregenomic RNA is the template for the viral reverse transcriptase, which replicates new viral DNA. All four viral proteins are also made. Transient viral replication occured in the mouse liver for about 1 week. In addition, RNAi has achieved regression of clinical traits in neurodegenerative disease model [121] but its potential for use in pharmaceutical target validation and as a therapeutic tool is still ongoing.

The ability to induce RNAi across mucosal surfaces was also investigated as a means for treating de sexually transmitted disorders [122]. siRNA targeting tumor necrosis factor alpha was injected into the joints of mice with collagen induced arthritis (CIA) and the development of arthritis was scored by assessing the inflammation of joints in the mouse paw, and the mice with CIA, joint inflammation was successfully inhibited [123]. Antiviral RNAi therapeutics have already entered human clinical trials and will hopefully prove to be safe and efficacious.

4.3. RNAi and cancer

The discovery of RNAi led to the realization that the RNAi machinery is also involved in normal gene regulation through the action of a class of small RNAs known as microRNAs. There is experimental evidence that miRNAs regulate cell division, differentiation, cell fate decisions, development, oncogenesis, apoptosis, and many other processes [35]. miRNA levels are also dramatically shifted in various cancers, and miRNAs can act as oncogenes [35]. It is now clear that miRNAs represent a gene regulatory network of enormous significance. The expression profile of miRNA is highly specific for a particular type of tissue and cell stage of cell differentiation [124]. Impaired miRNA functioning, which occurs during tumor transformation, can be evaluated as a consequence rather than the cause of loss of cell identity. However, detection of chromosomal rearrangement like deletions, local amplifications and chromosomal breakpoints in the region of miRNA genes (causing impairments in miRNA expression during cancerogenesis) is a good demonstration of direct role of miRNA in these processes.

Defects in miRNA expression could cause the development of cancer associated with impaired formation of oncoproteins or tumor supressors regulated by these miRNAs. One of the first identified miRNAs, let-7 found in *C. elegans* as well as in humans, is a tumor suppressor decreasing formation of the oncoproteins Ras and HMGA2 (High Mobility Group protein A2). In patients with lung cancer was a inverse correlation between expression of let-7 and Ras/HMGA2 [125]. miR-155 is an exemple of miRNA exhibiting oncogenic properties. This miRNA is required for B- and T- cells functioning [126]. Increased expression of miR-155 was observed in paediatric Burkitt's lymphoma, diffuse large cell lymphoma, Hodgkin's disease, lung, breast and pancreatic cancers [124]. In some tumor cells miR-21 and mi-R 24 act as oncogenes, and in others they act as tumor supressors. In HeLa cells, inhibition of miR-21 or miR-24 activity by modified anti-miR oligonucleotides accelerated proliferation [127]. Analysis of bioinformatic predictions of putative targets suggests that proto- and anti-oncogenic activity is typical for various miRNA [128].

Defects in miRNA expression associated with carcinogens are caused not only by the chromosomal rearrangements, but also due to impairments in the machinery responsible for miRNA formation and processing. Inhibition of expression of ribonucleases Dicer and Drosha by complementary siRNA caused acceleration of growth of lung adenocarcinoma cells in murine model [129]. Tumor transformation may be also determined by primary impairments in regulation of expression of a single miRNA, which is then accompanied by imbalance in the entire miRNA network [124]. Impairments in miRNA functioning seen in cancerogenesis can be used for detection of miRNA expression for diagnostics of tumor origin. Each type of cancer is characterized by a certain pattern of miRNA expression [124]. Moreover, the evaluation of miRNA profiles can be used for prognosis of the development of tumors [130].

The importance of epigenomic modifications of chromatin structure for the development of tumors has been recognized [13]. Methylation of cytosine in DNA followed by formation of 5-methylCytozine occurs in dinucleotide sequence CpG. Methylation can cause suppression of transcription by proteins recognizing methylated CpG attraction. Excesive methylation of CpG island of miRNA genes has been found in cancer cells [131]. The authors attribute the effect of this methylation to suppression of tumor suppressor genes represented by miRNA genes. Processes of DNA methylation may be closely associated with modification of chromatin histones. Some histone modifications like acetylation, methylation and phosphorylation of specific aminoacides residues are involved in gene expression. Impairments of histone modifications in tumor cells have been found in many studies [132]. Processes of epigenomic silencing may involve another type of non-coding RNAs, short siRNAs. This observation demonstrates the existence of nuclear RNA-interference, which is based on suppression of mRNA translation. Some experimental data suggest that siRNA plays a certain role in gene silencing at the level of chromatin in human cancer cells. There are some examples of involvement of short RNAs in epigenomic silencing, which is coupled to DNA methylation, histone modification of target genes, and attraction of the heterochromatin HP1 protein to them. All these chromatin modifications typically occur in the cancer epigenome [124]. The role of RNAi is recognized not only in silencing of proto-oncogenes or tumor suppressors, but also in maintenance of heterochromatin structure of centromeric region in mammalian cells [133].

4.4. Challenges for RNAi as a tool for diseases inverstigations

One of the advantages of RNAi over gene knockout is the ability to restrict gene knockdown to specific tissues or even cell types. This is important when a disease is a result of a mutation in an essential gene. The versatility of the technique has led to many applications. RNAi can be used in drug target validation and RNAi can target specific spliced exons, enabling the investigation of the functional roles of alternatively spliced forms of a gene [134]. An important opportunity is the use of RNAi in identification of all candidate genes involved in certain physiological processes using genome-wide RNAi screening [135].

RNAi can be applied to genetic model organisms such *Drosophila*, *C. elegans* and mouse in order to investigate and/or to treat some human disorders. Several models of human neural and neuromuscular disorders are available in this three experimental models. *C. elegans* is model for RNAi knockdown of genes to mimic loss of in order to elucidate the mechanisms of a number of muscle wasting diseases like: Duchenne muscular dystrophy [136], X-linked form of Emery-Dreifuss muscular distrophy [137], spinal muscular atrophy [137], fragile X-syndrome [138], Alzheimer's disease [139]. RNAi has been used in combination with overexpression studies to study the role of Parkin, an E3 ubiquitin ligase, in dopamine neuron degeneration in *Drosophila* in order to investigate the molecular mechanisms underlying Parkinson's disease. Overexpression of Parkin was shown to degrade its substrate (Pael-R) and suppress its toxicity, whereas interfering with endigenous Parkin promoted substrate accumulation and augmented its neurotoxicity [140].

Other experiment uses *Drosophila* in order to investigate human neurodegenerative disorders that are caused by expansion of CAG trinucleotide repeat. RNAi has been performed on two such diseases: Huntington's disease and spinobulbar muscular atrophy. *Drosophila* S2 cells were engineered to express a portion of human ar gene with CAG tracs of 26, 43 or 106 repeats tagged by green fluorescent protein (GFP) [141]. Cells carrying CAG repeats of 43 and 106 developed GFP aggregates. Using RNAi directed against AR protein, a loos of AR-GFP aggregates by 80% in co-transfected S2 cells has been observed. Therefore, RNAi could have considerable therapeutic potential in neurodegenerative disorders [134]. Murine model was used to investigate the potential of RNAi as therapeutic tool in neurodegenerative disorders. Spinocerebellar ataxia type 1 (SCA1) has been successfully suppressed by RNAi in mouse model of this disease [121].

RNAi might also allow future treatments of human disease such as Lou Gehrig's disease (amyotrophic lateral sclerosis, ALS), a genetically dominant inherited disease. This pattern of inheritance allowed identification of a specific gene that is linked to the death of Betz cells in some families, the gene for an enzyme called superoxide dismutase. Superoxide dismutase can protect cells from molecular damage caused by free radicals of oxygen. Mutant forms of superoxide dismutase can lead to cell death [142].

The Nobel Prize-winning research on RNA inhibition might lead to new treatments for patients with this disease due to dominant mutations in the superoxide dismutase gene. Reduction in the level of the superoxide dismutase enzyme coded for by the mutant gene has been studied in animal models [143]. Working with laboratory mice as an experimental model system for the human disease ALS, Miller and coworkers showed that loss of muscle function could be slowed using RNA interference [144]. This result was obtained by using a virus to induce RNA interference in neurons. These results from laboratory experiments suggest that if RNA-induced inhibition of mutant superoxide dismutase can be induced in the correct cells of the brain and spinal cord, it might be possible to slow progression of *Lou Gehrig's disease* in humans.

A high number of pharmaceutical and biotechnology companies have declared an interest in or have an active drug development program already underway in RNAi-based therapeutics to silence disease associated genes. This web (www.rnaiweb.com) collects together the latest research covering the development of RNAi based tools for drug target and gene function analysis. These include Sirna Therapeutics (Colorado) for macular degeneration; Avocel (Sunnyvale, California) for hepatits C; Alnylam Pharmaceuticals (Cambridge) for Parkinson's disease; CytRx (Los Angeles, California) for obesity, type II diabetes and ALS etc. But the major challenge in turning RNAi into an effective therapeutic strategy is the delivery of the RNAi agents, whether they are synthetic short double stranded RNAs or viral vectors directing production of double stranded RNA. During diseases, changes in the pattern of microRNAs will occur with some being indicative of treatment outcome and disease progression. More exciting then diagnostic value is the evidence that directly involves miRNAs in a number of diseases (cancer, imprinting impairments etc). In this context there is an increased interest in manipulations miRNAs for therapeutic purposes. In the loos of miRNA function, one approch is to mimic miRNA activity by introducing microRNA „mimics"with the same genetic information as the natural miRNA. For exemple, by adding more of a microRNA named let-t, it has been possible to halt cancer cells from further multiplying [145].

Another complementary approach to using miRNAs for therapy is to inhibit the activity of disease-associated miRNAs. This can be achieved by employing antisense oligonucleotides that, based on sequence complementarity, will bind to inactivate miRNA function. Esau end colleagues demonstrated that the inhibition of miRNA may be a potential therapeutic approach to the treatment of disease [146]. They inhibited miR-122 expression with antagomirs, which resulted in reduced plasma cholesterol levels and a decrease in hepatic fatty aceid and cholesterol synthesis rates in normal mice and in diet-induced obese mice. Targeting miR-122 with antagomirs resulted in inhibition of disease development.

RNA interference has much promise in laboratory. In principle, RNAi might be used to treat any disease that is linked to expression of an identified gene [112]. The most important challenge in turning RNA interference into an effective therapeutic strategy is the delivery of the RNA interference agents. Given sufficient research into delivery methods, some of these diseases will probably be treated effectively by RNAi based therapeutics.

5. Conclusions

The study of RNAi has led to a revolution in the understanding of gene expression and the examples in plants, animal and mammalian system as reviewed here, showed the diversity and the potential of RNAi as new approach to replaces the classical genetic technologies and manipulation.

Today, the RNAi strategies as a new tool for cheap screen of gene function in organisms for which, a genetic approach was not developed yet. However, after 11 years of extensive research, RNAi has now been demonstrated to function in mammalian cells to alter gene expression and used as a means for genetic discovery as well as a possible strategy for genetic correction and genetic therapy in cancer and other diseases.

Finally, RNAi represent a significant tool for the accomplishment of these goals, and will undoubtedly be used to address many other challenges in eukaryote functional genomics.

In future, a combination of RNAi and whole genome sequencing can contribute to the enhancement of the drug development success rates through better targets and RNAi platform efficiencies and keeping waste to a minimum by only treating people genetically predicted to respond to the therapeutic.

Author details

Gabriela N. Tenea and Liliana Burlibasa
Department of Genetics, University of Bucharest, Romania

Acknowledgement

GNT and LB were supported by The National Council of the Higher Education's Scientific Research (CNCSIS), IDEAS project: PNII 1958/2009.

6. References

[1]   Cogoni C, Macino G (2000) Post-transcriptional gene silencing across kingdoms. *Current Opinion in Genetic Development* 10(6): 638–643.

[2]   Mello CC, Darryl Conte Jr (2004) Revealing the world of RNA interference. *Nature* 431: 338-342.

[3]   Napoli C, Lemieux C, Jorgensen R (1990) Introduction of a chimeric chalcone synthase gene into petunia results in reversible co-suppression of homologous genes in trans. *Plant Cell* 2(4): 279–289.

[4]   De Haan P, Gielen JJL, Prins M, Wijkamp IG, Van Schepen A, Peters D, Van Grinsven MQJM, Goldbach R (1992) Characterization of RNA-mediated resistance to tomato spotted wilt virus in transgenic plants. *Bio/Technology* 10: 1133-1137.

[5]   Lindbo JA, Dougherty WG (1992) Untranslatable transcripts of the tobacco etch virus coat protein gene sequence can interfere with tobacco etch virus replication in transgenic plants and protoplasts. *Virology* 189: 725-733.

[6] van der Vlugt RA, Ruiter RK, Goldbach R (1992) Evidence for sense RNA-mediated protection to PVY[N] in tobacco plants transformed with the viral coat protein cistron. *Plant Molecular Biology* 20: 631-639.

[7] Ratcliff FG, MacFarlane SA, Baulcombe DC (1999) Gene silencing without DNA. RNA-mediated cross-protection between viruses. *Plant Cell* 11: 1207-1216.

[8] Zamore PD (2002) Ancient pathways programmed by small RNAs. Science 296: 1265-1269.

[9] Ruvkun G (2001) Glimpses of a tiny RNA world. *Science* 294 (5543): 797-799.

[10] Fire A, Xu S, Montgomery M, Kostas S, Driver S, Mello C (1998) Potent and specific genetic interference by double-stranded RNA in *Caenorhabditis elegans*. *Nature* 391: 806-811.

[11] Ketting RF, Plasterk RH, (2000) A genetic link between suppression and RNA interference in *C. elegans*. *Nature* 404: 296-298.

[12] Elbashir SM, Lendeckel W, Tuschl T (2001a) RNA interference is mediated by 21- and 22-nucleotide RNAs. *Genes and Development* 15(2): 188–200.

[13] Bernstein BE, Meissner A, Lander ES (2007) The mammalian epigenome. *Cell* 128: 669-681.

[14] Wargelius A, Ellingsen S, Jose FA (1999) Double stranded RNA induces specific developmental defects in zebrafish embryos. *Biochemical and Biophysical Research Communications* 263(1):156-161.

[15] Lohmann JU, Endl I, Bosch TC (1999) Silencing of developmental genes in Hydra. *Developmental Biology* 214(1): 211-214.

[16] Akashi H, Miyagishi M, Taira K (2001) Suppression of gene expression by RNA interference in cultured plant cells. *Antisense Nucleic Acid Drug Development* 11(6):359-367.

[17] Elmayan T, Balzergue S, Beon F, Bourdon V, Daubremet J, Guenet Y, Mourrain P, Palauqui JC, Vernhettes S, Vialle T, Wostrikoff K, Vaucheret H (1998) *Arabidopsis* mutants impaired in cosuppression. *Plant Cell* 10(10):1747-1758.

[18] Thakur A (2003) RNA interference revolution. *Electronic Journal Of Biotechnology* 6 (1).

[19] Wianny F, Zernicka-Goetz M (2000) Specific interference with gene function by double-stranded RNA in early mouse development. *Nature Cell Biology* 2: 70-75.

[20] Kreutzer R, Limmer S (2000) Kreutzer-Limmer Patent(1999/2000), Germany.

[21] Elbashir SM, Harborth J, Lendeckel W, Yalcin A, Weber K, Tuschl T (2001b) Duplexes of 21-nucleotide RNAs mediate RNA interference in cultured mammalian cells. *Nature* 411: 494-498.

[22] Watson JM, Fusaro FM, Wang M, Waterhouse PM (2005) RNA silencing platforms in plants. *Febs Letters* 579: 5982-5987.

[23] Matzke MA, Matzke AJ, Pruss GJ, Vance VB (2001) RNA-based silencing strategy in plants. *Current Opinion In Genetic Development* 11: 221-227.

[24] Tabara H, Grishok A, Mello CC (1998) RNAi in *C. elegans*: soaking in the genome sequence. *Science* 282 (5388): 430–431.

[25] Waterhouse PM, Wang MB, Lough T (2001) Gene silencing as an adaptive defence against viruses. *Nature* 411: 834-842.

[26] Meister G, Tuschl T (2004) Mechanisms of gene silencing by double-stranded RNA. *Nature* 431: 343-349.

[27] Kamath RS, Martinez-Campos M, Zipperlen P, Fraser AG, Ahringer J (2001) Effectiveness of specific RNA-mediated interference through ingested double-stranded RNA in *Caenorhabditis elegans*. *Genome Biology* 2(1): 2.1–2.10.

[28] Waterhouse PM, Helliwell CA, (2003) Exploring plant genomes by RNA induced gene silencing. *Nature Reviews: Genetics* 4: 29-38.

[29] Agrawal NP, Dasaradhi VN, Asif M, Pawan M, Bhatnagar RK, Mukherjee SK, (2003) RNA interference: biology mechanism and applications. *Microbiology and Molecular Biology Reviews* 67: 4 657–685.

[30] Baulcombe D (2004) RNA silencing in plants. *Nature* 43: 356-363.

[31] Hammond SM, Bernstein E, Beach D, Hannon GJ (2000) An RNA-directed nuclease mediates post-transcriptional gene silencing in *Drosophila* cells. *Nature* 404 (6775): 293–296.

[32] Hamilton AJ, Baulcombe DC (1999) A species of small antisense RNAi posttranscriptional gene silencing in plants. *Science* 286: 950-952.

[33] Zamore PD, Tuschl T, Sharp PA, Bartel DP (2000) RNAi: double-stranded RNA directs the ATP-dependent cleavage of mRNA at 21 to 23 nucleotide intervals. *Cell* 101(1): 25–33.

[34] Voinnet O, Lederer C, Baulcombe DC (2000) A viral movement protein prevents spread of the gene silencing signal in *Nicotiana benthamiana Cell* 103: 157-167.

[35] Rhoades MW, Reinhart BJ, Lim LP, Burge CB, Bartel B, Bartel DP (2002) Prediction of plant microRNA targets. *Cell* 110: 513-520.

[36] Fjose A, Drivenes O (2006) RNAi and microRNAs: from animal models to disease therapy. *Birth Defects Research (part C)* 78: 150-171.

[37] Okamura K, Ishizuka A, Siomi H, Siomi MC (2004) Distinct roles for argonaute proteins in small RNA-directed cleavage pathways. *Genes and Development* 18: 1655-1666.

[38] Gregory RI, Chendrimada TP, Cooch N, Shiekhattar R (2005) Human RISC couples microRNA biogenesis and posttranscriptional gene silencing. *Cell* 123: 631-640.

[39] Kim DH, Behlke MA, Rose SD, Chang MS, Choi S, Ross J (2005) Synthetic dsRNA Dicer substrates enhance RNAi potency and efficacy. *Nature Biotechnology* 23: 222-226.

[40] Siolas D, Lerner C, Burchard J, Ge W, Linsley PS, Paddison PJ, Hannon GJ, Cleary MA (2005) Synthetic shRNAs as potent RNAi triggers. *Nature Biotechnology* 23: 227-231.

[41] Doench JS, Petersen CP, Sharp PA (2003) siRNAs can function as miRNAs. *Genes Development* 17: 438-442.

[42] Miska EA (2005) How microRNAs control cell division differentiation and death. *Current Opinion Genetic Development* 15: 563-568.

[43] Alvarez-Garcia I, Miska EA, (2005) MicroRNA functions in animal development and human disease. *Development* 132: 4653-4662.

[44] Schiebel W, Pelissier T, Riedel L, Thalmeir S, Schiebel R, Kempe D, Lottspeich F, Sanger HL, Wassenegger M (1998) Isolation of an RNA-directed RNA polymerase-specific cDNA clone from tomato. *Plant Cell* 10: 2087-2101.

[45] Cullen BR (2005) RNAi the natural way. *Nature Genetics* 37:1163-1165.

[46] Krutzfeldt J, Rajewsky N, Braich R, Rajeev KG, Tuschl T, Manoharan M, Stoffel M (2005) Silencing of microRNAs in vivo with „antagomirs". *Nature* 438: 685-689.

[47] Metzalapff M (2005) Applications of RNAi in crop improvement *Pflanzenschutz-Nachrichten Bayer* 58(1): 51-59.

[48] Rodriguez A, Griffiths-Jones S, Ashurst JL (2004) Identification of mammalian microRNA host genes and transcription units. *Genome Research* 14: 1902-1910.

[49] Tang G, Galili G (2004) Using RNAi to improve plant nutritional value: from mechanisms to application. *Trends In Biotechnology* 22(9): 463-469.

[50] Ying SY, Lin SL (2004) Intron derived microRNAs-fine tunning of gene functions. *Gene* 342: 25-28.

[51] Zhao Y, Srivastava D (2007) A developmental view of microRNA function. *Trends Biochemistry Science* 32(4): 189-97.

[52] Lee RC, Feinbaum RL, Ambros V (1993) The *C. elegans* heterochronic gene lin-4 encodes small RNAs with antisense complementary to lin-14. *Cell* 75: 843-854.

[53] Lagos-Quintana M, Rauhut R, Lendeckel W (2001) Identification of novel genes coding for small express RNAs. *Science* 294: 853-858.

[54] Lee RC, Ambros V (2001) An extensive class of small RNAs in *Caenorhabditis elegans*. *Science* 294: 862-864.

[55] Ketting RF, Fischer SEJ, Bernstein E, Sijen T, Hannon GJ, Plasterk RHA (2001) Dicer functions in RNA interference and in synthesis of small RNA involved in developmental timing in *C. elegans. Genes and Development* 15: 2654-2659.

[56] Knight SW, Bass BL (2001) A role for the RNAse III enzyme DCR-1 in RNA interference and germ line development in *Caenorhabditis elegans. Science* 293: 2269-2271.

[57] Yekta S, Shih IH, Bartel DP (2004) MicroRNA-directed cleavage of HOXB8 mRNA. *Science* 304: 594-596.

[58] Yang M, Mattes J (2008) Discovery biology and therapeutic potential of RNA interference microRNA and antagomirs. *Pharmacology and Therapeutics* 117: 94-104.

[59] Baskerville S, Bartel DP (2005) Microarray profiling of microRNAs reveals frequent coexpression with neighboring miRNAs and host genes. *RNA* 11: 241-247.

[60] Du TT, Zamore PD (2005) microPrimer: the biogenesis and function of microRNA. *Development* 132: 4645-4652.

[61] Jones-Rhoades MW, Bartel DP (2004) Computational identification of plant microRNAs and their targets including a stress-induced miRNA. *Molecular Cell* 14: 787-799.

[62] Llave C, Kasschau KD, Rector MA, Carrington JC (2002) Endogenous and silencing-associated small RNAs in plants. *Plant Cell* 14: 1605-1619.

[63] Park W, Li J, Song R, Messing J, Chen X (2002) CARPEL FACTORY: a Dicer homolog and HEN1 a novel protein act in microRNA metabolism in *Arabidopsis thaliana. Current Biology* 12: 1484-1495.

[64] Peragine A, Yoshikawa M, Wu G, Albrecht HL, Poethig RS (2004) *SGS3* and *SGS2/SDE1/RDR6* are required for juvenile development and the production of *trans*-acting siRNAs in *Arabidopsis. Genes and Development* 18: 2368–2379.

[65] Vazquez F, Vaucheret H, Rajagopalan R, Lepers C, Gasciolli V, Mallory AC, Hilbert JL, Bartel DP, Crete P (2004) Endogenous trans-acting siRNAs regulate the accumulation of *Arabidopsis* mRNAs. *Molecular Cell* 16: 69–79.

[66] Aravin AA, Naumova NM, Tulin AV, Vagin VV, Rozovsky YM, Gvozdev VA (2001) Double-stranded RNA-mediated silencing of genomic tandem repeats and transposable elements in the *D. melanogaster* germline. *Current Biololy* 11: 1017–1027.

[67] Reinhart BJ, Bartel DP (2002) Small RNAs correspond to centromere heterochromatic repeats. *Science* 297: 1831.

[68] Volpe TA, Kidner C, Hall IM, Teng G, Grewal SIS, Martienssed RA (2002) Regulation of heterochromatic silencing and histone H3 lysine-9 methylation by RNAi. *Science* 297: 1833–1837.

[69] Mochizuki K, Gorovsky MA (2004) Small RNAs in genome rearrangement in *Tetrahymena. Current Opinion in Genetic Development* 14: 181–187.

[70] Aravin A, Pfeffer GD, Lagos-Quintana M, Landgraf P, Iovino N, Morris P, Brownstein MJ, Kuramochi-Miyagawa S, Nakano T, Chien M, Russo JJ, Sheridan R, Sander C, Zavolan M, Tuschl T (2006) A novel class of small RNAs bind to MILI protein in mouse testes. *Nature* 442: 203–207.

[71] Catalanotto C, Azzalin G, Macino G, Cogoni C (2000) Gene silencing in worms and fungi. *Nature* 404: 245.

[72] Tabara H, Sarkissian M, Kelly WG, Fleenor J, Grishok A, Timmons L, Fire A, Mello CC (1999) The RDE-1 gene RNA interference and transposon silencing in *C. elegans. Cell* 99: 123–132.

[73] Xie Z, Johansen LK, Gustafson AM, Kasschau KD, Lellis AD, Zilberman D, Jacobsen SE, Carrington JC (2004) Genetic and functional diversification of small RNA pathways in plants. *Plos Biology* 2: E104.

[74] Bohmert K, Camus I, Bellini C, Bouchez D, Caboche M, Benning C (1998) AGO1 defines a novel locus of *Arabidopsis* controlling leaf development. *EMBO Journal* 17: 170-180.

[75] Vaucheret H, Vasquez F, Crete P, Bartel DP (2004) The action of ARGONAUTE1 in the miRNA pathway and its regulation by the miRNA pathway are crucial for plant development. *Genes and Development* 18: 1187-1197.

[76] Zilberman D, Cao X, Jacobsen SE (2003) ARGONAUTE 4 control of locus-specific siRNA accumulation and DNA and histone methylation. *Science* 299: 716-719.

[77] Kosik KS (2006) The neuronal microRNA system *Nature Reviews. Neurosciences* 7: 911-920.

[78] Dalmay T, Hamilton A, Rudd S, Angell S, Baulcombe DC, (2000) An RNA-dependent RNA polymerase gene in *Arabidopsis* is required for posttranscriptional gene silencing mediated by a transgene but not by a virus. *Cell* 101: 543-553.

[79] Cogoni C, Macino G (1999) Gene silencing in *Neurospora crassa* requires a protein homologous to RNAdependent RNA polymerase. *Nature* 399: 166-169.

[80] Smardon A, Spoerke JM, Stacey SC, Klein ME, Mackin N, Maine EM (2000) EGO-1 is related to RNA-directed RNA polymerase and functions in germ-line development and RNA interference in *C. elegans. Current Biology* 10: 169-178.

[81] Wu-Scharf D, Jeong B, Zhang C, Cerutti H (2000) Transgene and transposon silencing in *Chlamydomonas reinhardtii* by a DEAH-box RNA helicase. *Science* 290: 1159-1162.

[82] Hammond SM (2005) Dicing and slicing: the core machinery of the RNA interference pathway. *FEBS Letters* 579: 5822-5829.

[83] Sato F (2005) RNAi and functional genomics. *Plant Biotechnology* 22: 431-442.

[84] Tenea GN (2009) Exploring the world of RNA interference in plant functional genomics: a research tool for many biology phenomena. *Roumanian Biotechnological Letters* 14(3) 4360-4364.

[85] Jones L, Ratcliff F, Baulcombe DC, (2001) RNA-directed transcriptional gene silencing in plants can be inherited independently of the RNA trigger and requires Met1 for maintenance. *Current Biology* 11: 747–757.

[86] Matthew L (2004) RNAi for plant functional genomics. *Comparative And Functional Genomics* 5: 240-244.

[87] Hilson P, Allemeersch J, Altmann T, Aubourg S, Avon A, Beynon J, Bhalerao RP, Bitto F, Caboche M, Cannoot B, Chardakov V, Cognet-Holliger C, Colot V, Crowe M, Darimont C, Durinck S, Eickhoff H, Falcon De Longevialle A, Farmer EE, Grant M, Kuiper MTR, Lehrach H, Léon C, Leyva A, Lundeberg J, Lurin C, Moreau Y, Nietfeld W, Paz-Ares J, Reymond P, Rouzé P, Sandberg G, Dolores Segura M, Serizet C, Tabrett A, Taconnat L, Thareau V, Van Hummelen P, Vercruysse S, Vuylsteke M, Weingartner M, Weisbeek PJ, Wirta V, Wittink FRA, Zabeau M, Small I (2004) Versatile gene-specific sequence tags for *Arabidopsis* functional genomics: Transcript profiling and reverse genetics applications. *Genome Research* 14: 2176-2189.

[88] Invitrogen Gateway^{tm} Technology. (2004) *Quest* 1(2): 32-33.

[89] Ifuku K, Yamamoto Y, Sato F (2003) Specific RNA interference in psbP genes encoded by multigene family in *Nicotiana tabacum* with a short 3′-untranslated sequence. *Bioscience Biotechnology Biochemistry* 67: 107-113.

[90] Miki D, Itoh R, Shimamoto K (2005) RNA silencing of single and multiple members in a gene family of rice. *Plant Physiology* 138: 1903-1913.

[91] Yamamoto Y, Ifuku K, Sato F (2005) Suppression of psbP and psbQ genes in *Nicotiana tabacum* by RNA interference technique. In: van der Est A. Bruce D. (eds) *Photosynthesis: Fundamental Aspects to Global Perspectives*. Kluwers Academic Publisher The Netherlands pp 798-799.

[92] Auer C, Frederick R (2009) Crop improvement using small RNAs: applications and predictive ecological risk assessments. *Trends Biotechnology* 27: 644-651.

[93] Gheysen G, Vanholme B (2007) RNAi from plants to nematodes. *Trends Biotechnology* 25: 89-92.

[94] Gordon KHJ, Waterhouse PM (2007) RNAi for insect-proof plants. *Nature Biotechnology* 25: 1231-1232.

[95] Sunilkumar G, Campbell LM, Puckhaber L, Stipanovic RD, Rathore KS (2006) Engineering cottonseed for use in human nutrition by tissue-specific reduction of toxic gossypol. *Proceedings of the National Academy of Sciences* 103: 18054-18059.

[96] Wang M, Abbott D, Waterhouse PM (2000) A single copy of a virus derived transgene encoding hairpin RNA gives immunity to barley yellow dwarf virus. *Molecular Plant Pathology* 1: 401-410.

[97] Kusaba M, Miyahara K, Lida S, Fukuoka H, Takario T, Sassa H, Mischimura M, Nischio T (2003) Low-glutein content 1: a dominant mutation that suppresses the glutein multigene family via RNA silencing in rice. *Plant Cell* 15: 1455-1467.

[98] Voelker T, Kinney AJ (2001) Variations in the biosynthesis of seed storage lipids. *Annual Review Plant Physiolology. Plant Molecular Biology* 52: 335–361.

[99] Knowlton S (1999) Soybean oil having high oxidative stability. US Patent 5981781.

[100] Mroczka A, Roberts PD, Fillatti JJ, Wiggins BE, Ulmasov T, Voelker T (2010) An intron sense suppression construct targeting soybean FAD2-1 requires a double-stranded RNA-producing inverted repeat T-DNA insert. *Plant Physiology* 153: 882–891.

[101] Hoffer P, Ivashuta S, Pontesc O, Vitins A, Pikaard C, Mroczka A, Wagnera N, Voelker T (2011) Posttranscriptional gene silencing in nuclei. *Proceedings of the National Academy of Sciences* USA 108(1): 409-414.

[102] Matthew L (2004) RNAi for plant functional genomics. *Comparative And Functional Genomics* 5: 240-244.

[103] Shi XM, Miller H, Verchot J, Carrington JC, Vance VB (1997) Mutation in the region encoding the central domain of helper component-proteinase (HC-Pro) eliminate potato virus X/potyviral synergism. *Virology* 231: 35-42.

[104] Pruss G, Ge X, Shi XM, Carrington JC, Bowman Vance V (1997) Plant viral synergism: the potyviral genome encodes broad–range pathogenicity enhancer that transactivates replication of heterologous viruses. *Plant Cell* 9: 859-868.

[105] Roth BM, Pruss GJ, Vance VB (2004) Plant viral suppressors of RNA silencing. *Virus Research* 102: 97-108.

[106] Moissiard G, Voinnet O (2004) Viral suppression of RNA silencing in plants. *Molecular Plant Phatology* 5: 71-82.

[107] Chapman EJ, Prokhnevsky AI, Gopinath K, Dolja V, Carrington JC (2004) Viral RNA silencing suppressors inhibits the microRNA pathways at an intermediate step. *Genes Development* 18: 1179-1186.

[108] Dunoyer P, Lecellier CH, Parizotto EA, Himber C, Voinnet O (2004) Probing the microRNA and small interfering RNA pathways with virus-encoded suppressors of RNA silencing. *Plant Cell* 16: 1235-1250.

[109] Anandalakshmi R, Pruss GJ, Ge X, Marathe R, Mallory AC, Smith TH, Vance VB (1998) A viral suppressor of gene silencing in plants. *Proceedings of the National Academy of Sciences* USA 95: 13079-13084.

[110] Kasschau KD, Carrington JC (1998) A counterdefensive strategy of plant viruses: suppression of posttranscriptional gene silencing. *Cell* 95: 461-470.

[111] Mallory AC, Mlotshwa S, Bowman LH, Vance VB (2003) The capacity of transgenic tobacco to send a systemic RNA silencing signal depends on the nature of the inducing transgene locus. *Plant Journal* 35: 82-92.

[112] Downward J (2004) RNA interference. *British Medical Journal* 328: 1245-1248.

[113] Jacque JM, Triques K, Stevenson M (2002) Modulation of HIV-1 replication by RNA interference. *Nature* 418: 435–438.

[114] Lee NS, Dohjima T, Bauer G, Li H, Ehsani A, Salvaterra P, Rossi J (2002) Expression of small interfering RNAs targeted against HIV-1 *rev* transcripts in human cells. *Nature Biotechnology* 20: 500–505.

[115] Novina CD, Murray MF, Dykxhoorn DM, Beresford PJ, Riess J, Lee SK, Collman RG, Lieberman J, Shankar P, Sharp PA (2002) siRNA-directed inhibition of HIV-1 infection. *Nature Medicine* 8: 681–686.

[116] Ananthalakshmi P, Sutton R (2008) Titers of HIV-based vectors encoding shRNAs are reduced by a Dicer-dependent mechanism. *Molecular Therapy* 16: 378-386.

[117] Yu JY, Deruiter SL, Turner DL (2002) RNA interference by expression of short interfering RNAs and hairpin RNAs in mammalian cells *Proceedings of the National Academy of Sciences* USA 99: 6047-6052.

[118] Hannon GJ, Rossi JJ (2004) Unlocking the potential of the human genome with RNA interference. *Nature* 431: 371-378.

[119] Chi JT, Chang HY, Wang NN, Chang DS, Dunthy N, Brown PO (2003) Genome wide view of gene silencing by small interfering RNAs. *Proceedings of the National Academy of Sciences* USA 100: 6343-6346.

[120] McCaffrey AP, Nakai H, Pandey K, Huang Z, Salazar FH, Xu H, Wieland SF, Marion PL, Kay MA (2003) Inhibition of hepatitis B virus in mice by RNA interference. *Nature Biotechnology* 21: 639-644.

[121] Xia H, Mao Q, Eliason S, Harper SQ, Martins IH, Orr HT, Laulson HL, Yang L, Kotin RM, Davidson BL (2004) RNAi suppresses polyglutamine-induced neurodegeneration in a model of spinocerebellar ataxia. *Nature Medicine* 10: 816-820.

[122] Palliser D, Chowdhury D, Wang Q, Lee SJ, Bronson RT, Knipe DM, Lieberman J (2006) An siRNA-based microbicide protects mice from lethal Herpex simplex virus 2 infection. *Nature* 439: 89-94.

[123] Schiffelers RM, Xu J, Storm G, Woodle MC, Scaria PV (2005) Effects of treatment with small interfering RNA on joint inflammation in mice with collagen-induced arthritis. *Arthritis Rheumatology* 52: 1314-1318.

[124] Ryazansky SS, Gvozdev VA (2008) Small RNAs and cencerogenesis. *Biokhimia* 73(5): 640-655.

[125] Johnson SM, Grosshans H, Shingara J, Byrom M, Jarvis R, Cheng A, Labourier E, Reinert KL, Brown D, Slack FJ (2005) RAS is regulated by the let-7 microRNA family. *Cell* 120: 635-647.

[126] O'Connell RM, Taganov KD, Boldin MP, Cheng G, Baltimore D (2007) MicroRNA-155 is induced during the macrophage inflammatory response. *Proceedings of the National Academy of Sciences* USA 104: 1604-1609.

[127] Cheng AM, Byrom MW, Shelton J, Ford L (2005) Antisense inhibition of human miRNAs and indications for an involvement of miRNA in cell growth and apoptosis. *Nucleic Acids Research* 33: 1290-1297.

[128] Janot G, Simard MJ (2006) Tumour-related microRNAs functions in *Caenorhabditis elegans*. *Oncogene* 25: 6197-6201.

[129] Kumar MS, Lu J, Mercer KL, Golub TR, Jacks T (2007) Impaired microRNA processing enhances cellular transformation and tumorigenesis. *Nature Genetics* 39: 673-677.

[130] Calin GA, Ferracin M, Cimmino A, Di LG, Shimizu M, Wojcik SE, Iorio MV, Visone R, Sever NI, Fabbri M, Iuliano R, Palumbo T, Picchiorri F, Roldo C, Garzon R, Sevignani C, Rassenti L, Alder H, Volinia S, Liu CG, Kipps TJ, Negrini M, Croce CM (2005) A MicroRNA signature associated with prognosis and progression in chronic lymphocytic leukemia. *The New England Journal of Medicine* 353: 1793-1801.

[131] Saito Y, Liang G, Egger G, Friedman JM, Chuang JC, Coetzee GA, Jones PA (2006) Specific activation of microRNA-127 with downregulation of the proto-oncogene BCL6 by chromatin-modifying drugs in human cancer cells. *Cancer Cell* 9: 435-443.

[132] Esteller M (2007) Cancer epigenomics: DNA methylomes and histone-modification maps. *Nature Review Genet*ics 8: 286-298.

[133] Kanellopoulou C, Muljo SA, Kung AL, Ganesan S, Drapkin R, Jenuwein T, Livingston DM, Rajewsky K (2005) Dicer deficient mouse embryonic stem cell are defective in differentiation and centromeric silencing. *Genes and Development* 19: 489-501.

[134] Buckingham SD, Esmaeili B, Wood M, Sattelle DB (2004) RNA interference: from model organisms towards therapy for neural and neuromuscular disorders. *Human Molecular Genetics* 13: R275-R288.

[135] Kamath RS, Ahringer J (2003) Genome-wide RNAi screening in *Caenorhabditis elegans*. *Methods* 30: 313-321.

[136] Eagle M, Boudouin SV, Chandler C, Giddings DR, Bullock R, Bushby K (2002) Survival in Duchenne muscular dystrophy: improvements in life expectancy since 1967 and the impact of home nocturnal ventilation. *Neuromuscular Disorders* 12: 926-929.

[137] Goldman RD, Gruenbaum Y, Moir RD, Shumaker DK, Spann TP (2002) Nuclear lamins: building blocks of nuclear arhitecture. *Genes and Development* 16: 533-547.

[138] Carthew RW (2002) RNA interference: the fragile X syndrome connection. *Current Biology* 12: R852-R854.

[139] Nollen EA, Garcia SM, van Haaften G, Kim S, Chavez A, Morimoto RI, Plasterk RH (2004) Genome-wide RNA interference screen identifies previously undescribed regulators of polyglutamine aggregation. *Proceedings of the National Academy of Sciences* USA 101: 6403-6408.

[140] Yang Y, Nishimura I, Imai Y, Takahashi R, Lu B (2003) Parkin suppresses dopaminergic neuron-selective neutotoxicuty induced by Pael-R in *Drosophila*. *Neuron* 37: 911-924.

[141] Caplen NJ, Taylor JP, Statham VS, Tanaka F, Fire A, Morgan RA (2002) Rescue of polyglutamine-mediated cytotoxicity by double-stranded RNA-mediated RNA interference. *Human Molecular Genetics* 11:175-184.

[142] Kononenko NI, Shao LR, Dudek FE (2004) Riluzole-sensitive slowly inactivating sodium current in rat suprachiasmatic nucleus neurons. *Journal of Neurophysiology* 91: 710-718.

[143] Smith RA, Miller TM, Yamanaka K, Monia BP, Condon TP, Hung G, Lobsiger CS, Ward CM, Wei H, Wancewicz Bennett CF, Cleveland DW (2006) Antisense oligonucleotide therapy for neurodegenerative disease*The Journal of Clinical Investigation* 116: 2290-2296.

[144] Miller M, Kaspar KP, Kops GJ, Yamanaka K, Christian LJ, Gage FH, Cleveland DW (2006) Virus–delivered small RNA silencing sustain strength in amyotrophic lateral sclerosis. *Annual Neurology* 57: 773-776.

[145] de Fougerolles A, Vornlocher HP, Maraganore J, Lieberman J (2007) Interfering with disease: a progress report on siRNA-based therapeutics. *Nature Reviews Drug Discovery* 6: 443-453.

[146] Esau C, Davis S, Murray SF, Yu XX, Pandey SK, Pear M, Watts L, Booten S.L, Graham M (2006) miR-122 regulation of lipid metabolism revealed by in vivo antisense targeting. *Cell Metabolism* 3: 87-98

Chapter 2

Analysis of Gene Expression Data Using Biclustering Algorithms

Fadhl M. Al-Akwaa

Additional information is available at the end of the chapter

1. Introduction

One of the main research areas of bioinformatics is functional genomics; which focuses on the interactions and functions of each gene and its products (mRNA, protein) through the whole genome (the entire genetics sequences encoded in the DNA and responsible for the hereditary information). In order to identify the functions of certain gene, we should able to capture the gene expressions which describe how the genetic information converted to a functional gene product through the transcription and translation processes. Functional genomics uses microarray technology to measure the genes expressions levels under certain conditions and environmental limitations. In the last few years, microarray has become a central tool in biological research. Consequently, the corresponding data analysis becomes one of the important work disciplines in bioinformatics. The analysis of microarray data poses a large number of exploratory statistical aspects including **clustering** and **biclustering** algorithms, which help to identify similar patterns in gene expression data and group genes and conditions in to subsets that share biological significance.

1.1. What is Clustering?

A large number of clustering definitions can be found in the literature. The simplest definition is shared among all and includes one fundamental concept: the grouping together of similar data items into clusters[1].

Clustering is an important explorative statistical analysis of gene expression data. It aims to identify and group genes that exhibit similar expression patterns over several conditions and also group the conditions based on the expression profiles across set of genes. The successful clustering approach should guarantee two criteria which are homogeneity high similarity between elements in the same cluster, and separation – low similarity between elements from different clusters. When homogeneity and separation are precisely defined, those are two opposing objectives: The better the homogeneity the poorer the separation, and vice versa [2]. Several algorithmic techniques were previously used for clustering gene expression data, including hierarchical clustering [3], self organizing maps [4], and graph theoretic approaches [5].

1.1.1. K-means

K-means is a classical clustering algorithm [6] invented in 1956 to classify or to group objects (genes) based on attributes or features (experimental conditions) into K number of groups (clusters). K is positive integer number and assumed to be known.

K-means computational approach starts by placing K points into the space represented by the objects that are being clustered. These points represent initial group centroids. We can take any random objects as the initial centroids or the first K objects in sequence can also be used as the initial centroids. Then the K means algorithm will do the four steps below until convergence:

1.   Determine the centroids coordinate.
2.   Determine the distance of each object to the centroids using the Euclidean distance.
3.   Group the objects based on minimum distance.
4.   Iterate the above steps till no object moves its assigned group.

Each iteration of k-means modifies the current partition by checking all possible modifications of the solution, in which one element is moved to another cluster. This is done by reducing the sum of distances between objects and the centers of their clusters. This procedure is repeated until no further improvement is achieved (No object move the group) and all the objects are grouped into the final required number of clusters.

A disadvantage of K-means algorithm could be perceived in the need to specify the number of clusters K as a parameter value prior to running the algorithm. In cases where there is no expectation about K, user has to make trails with several values of K or use external techniques to guess the no of clusters may be exist.

1.1.2. Hierarchical clustering (HCL)

Hierarchical clustering does not partition the genes into subsets. Instead it builds a down-top hierarchy of clusters using agglomerative methods or top - down hierarchy of clusters using divisive methods. The traditional graphical representation of this hierarchy is called dendrogram tree. The divisive method begins at the root and starts to breaks up clusters whose having low similarity. Whereas, the Agglomerative method begins at the leaves of the tree and starts with an initial partition into single element clusters and successively merges clusters until all elements belong to the same cluster [3]. (See Figure 1) The agglomerative method is widely used than the divisive one which is not generally available, and rarely has been applied. The idea of the agglomerative method can be summarized as following: Given a set of N items (genes in our case) to be clustered, and an N*N distance (or similarity) matrix [7],

1.   Assign each item to a cluster, so you have N clusters, each containing just one item.
2.   Find the closest (most similar) pair of clusters and merge them into a single cluster.
3.   Compute distances (similarities) between the new cluster and each of the old clusters.
4.   Repeat steps 2 and 3 until all items are clustered into a single cluster of size N.

In Step 3, distance or similarity measurements between the merged clusters and all the other clusters can be calculated in one of three schemes: single-linkage, complete linkage and average-linkage.

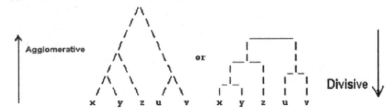

Figure 1. HCL: Agglomerative and Divisive Methods.

1.2. Biclustering

Traditional clustering approaches such as k-means and hierarchical clustering put each gene in exactly one cluster based on the assumption that all genes behave similarly in all conditions. However, recent understanding of cellular processes shows that it is possible for subset of genes to be co expressed under certain experimental conditions, and at the same time; to behave almost independently under other conditions. From this context, a new two mode clustering approach called biclustering or co-clustering has been introduced to group the genes and conditions in both dimensions simultaneously.

This allows finding subgroups of genes that show the same response under a subset of conditions, not all conditions. Also, genes may participate in more than one function, resulting in one regulation pattern in one context and a different pattern in another.

Example, if a cellular process is only active under specific conditions and there is a gene participates in multiple pathways that are differentially regulated, one would expect this gene to be included in more than one cluster; and this cannot be achieved by traditional clustering techniques.

Many biclustering methods exist in the literature [8]. Table 1 summarized some of promising biclustering algorithms developed during the last ten years. In brief, we described some of these algorithms according to their prediction strength, their promising results, to what they extend in the community, whether an implementation was available, and the feedback from their authors to explain some ambiguous issues.

1.2.1. Cheng and Church (CC)

CC algorithm[18] is considered to be the first real biclustering implementation after the primary idea has been introduced by Hartigan [19] in 1972.

Algorithm	Approach Time	Complicity	Prediction ability
Bivisu [9]	Exhaustive Bicluster Enumeration	$O(m^2 n \log m)$[a]	Coherent values
MSBE [10]	Greedy Iterative Search	$O((n+m)^2)$	Coherent values
Bimax[11]	Divide-and-Conquer	$O(nm\beta\log\beta)$	Coherent values
ROBA [12]	Matrix algebra	$O(nmLN)$	Coherent Evolution
x-motif [13]	Greedy Iterative Search	$nm^{O(\log(1/a)/\log(1/\beta))}$	Coherent Evolution
SAMBA [14]	Exhaustive Bicluster Enumeration	$O(n2)$	Coherent Evolution
OPSM [15]	Greedy Iterative Search	$O(nm^3 I)$	Coherent Evolution
Plaid[16]	Distribution Parameter Identification	XXX[b]	Coherent values
ISA [17]	Iterative Signature Algorithm	XXX	Coherent values
CC [18]	Greedy Iterative Search	$O((n+m)nm)$	Coherent values

[a] n and m are the row and column sizes of the expression matrix
[b] not available

Table 1. Biclustering Algorithms Comparison.

CC defines a bicluster as a subset of rows and a subset of columns with a high similarity. The proposed similarity score is called mean squared residue (H) and it is used to measure the coherence of the rows and columns in the single bicluster. Given the gene expression data matrix A = (X;Y); a bicluster is defined as a uniform submatrix (I;J) having a low mean squared residue score as following:

The CC Mean Squared Residue:

$$H(I,J) = \frac{1}{|I||J|}\sum_{i\in I, j\in J}\left(a_{ij} - a_{iJ} - a_{Ij} + a_{IJ}\right)^2$$

Where: a_{ij} is gene expression level at row i and column j, a_{iJ} is the mean of row i, a_{Ij} is the mean of column j, a_{IJ} is the overall mean. CC algorithm will identify the submatrix as a bicluster if the score is below a level alpha which is a user input parameter to control the quality of the output biclusters. Generally; CC algorithm performs the following major steps:

1. Delete rows and columns with a score larger than alpha.
2. Adding rows or columns until alpha level is reached.
3. Iterate these steps until a maximum number of biclusters is reached or no bicluster is found [18].

1.2.2. Iterative Signature Algorithm (ISA)

The ISA algorithm [17, 20] is a novel method for the biclustering analysis of large-scale expression data. It is an efficient algorithm based on the iterative application of the signature algorithm presented in [17]. ISA considers a bicluster to be a transcription module which can be defined as a set of coexpressed genes together with the associated set of regulating conditions (Figure 2). Starting with an initial set of genes, all samples (conditions) are scored with respect to this gene set and those samples are chosen for which the score exceeds a certain threshold (usually defined by the user). In the same way, all genes are scored regarding the selected samples and a new set of genes is selected based on another user-defined threshold. The entire procedure is repeated until the set of genes and the set of samples converge and do not change anymore.

Multiple biclusters can be discovered by running the ISA algorithm on several initial gene sets. This approach requires identification of a reference gene set which needs to be carefully selected for good quality results. In the absence of pre-specified reference gene set, random set of genes is selected at the cost of results quality[17].

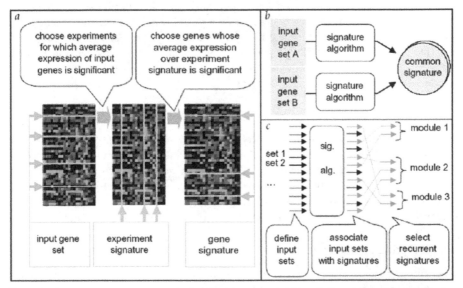

Figure 2. The recurrence signature method. a, The signature algorithm. b, Recurrence as a reliability measure. The signature algorithm is applied to distinct input sets containing different subsets of the postulated transcription module. If the different input sets give rise to the same module, it is considered reliable. c, General application of the recurrent signature method. Copyright © [17].

1.2.3. Biclusters Inclusion Maximal (Bimax)

Bimax[11] is a simple binary model and new fast divide-and-conquer algorithm used to cluster the gene expression data. It is presented in 2006 by Computer Engineering and Networks Laboratory ETH Zurich, Switzerland. Bimax discretized the gene expression data matrix and convert it into a binary matrix by identifying a threshold, so transcription levels (genes expression values) above this threshold become ones and transcription levels below become zeros (or vice versa). Then, it searches for all possible biclusters that contain only ones. This can be done by iterating these steps:

1.   Rearrange the rows and columns to concentrate ones in the upper right of the matrix.
2.   Divide the matrix into two sub matrices.
3.   Whenever in one of the submatrices only ones are found, this sub matrix is returned.

1.2.4. Order Preserving Submatrix(OPSM)

The order-preserving submatrix (OPSM) algorithm [15] is a probabilistic model introduced to discover a subset of genes identically ordered among a subset of conditions. It focuses on the coherence of the relative order of the conditions rather than the coherence of actual expression levels. In other words, the expression values of the genes within a bicluster induce an identical linear ordering across the selected conditions. Accordingly, the authors define a bicluster as a subset of rows whose values induce a linear order across a subset of the columns. The time complexity of this model is $O(nm^3I)$ where n andmare the number of rows and columns of the input gene expression matrix respectively and I is the number of biclusters. A disadvantage of OPSM algorithm is that it takes long time for high dimensional datasets. And this is because its time complexity is cubic with regards to the number of columns (dimensions) of the input matrix [15].

1.2.5. Maximum Similarity Bicluster(MSBE)

MSBE Biclustering algorithm [10] is a novel polynomial time algorithm to find an optimal biclusters with the maximum similarity. The idea behind this algorithm is to find subset of genes that are related to a reference gene. The reference gene is known in advance. MSBE algorithm uses the similarity score for a sub-matrix to find the similar expressions in the microarray datasets. And the threshold of the average similarity score is a user input parameter in order to allow the user to control the quality of the biclustering results.

1.3. Clustering or biclustering

Clustering algorithms [21-23] have been used to analyze gene expression data, on the basis that genes showing similar expression patterns can be assumed to be co-regulated or part of the same regulatory pathway. Unfortunately, this is not always true. Two limitations obstruct the use of clustering algorithms with microarray data. First, all conditions are given equal weights in the computation of gene similarity; in fact, most conditions do not contribute information but instead increase the amount of background noise. Second, each gene is assigned to a single cluster, whereas in fact genes may participate in several functions and should thus be included in several clusters[24].

A new modified clustering approach to uncovering processes that are active over some but not all samples has emerged, which is called biclustering. A bicluster is defined as a subset of genes that exhibit compatible expression patterns over a subset of conditions [11].

During the last ten years, many biclustering algorithms have been proposed (see [8] for a survey), but the important questions are: which algorithm is better? And do some algorithms have advantages over others?

Recently Kevin *et al.*[25]proposed a semantic web algorithm to recommend the best algorithm based on user inputs like: is the dataset contain outliers, is it allowed to get overlapped clusters and the time to retrieve the biclusters.

Generally, comparing different biclustering algorithms is not straightforward as they differ in strategies, approaches, time complicity, number of parameters and prediction ability. In addition, they are strongly influenced by user selected parameter values. For these reasons, the quality of biclustering results is often considered more important than the required computation time. Although there are some analytical comparative studies to evaluate the traditional clustering algorithms[21-23], for biclustering; no such extensive comparison exist even after initial trails have been taken [11]. In the end, Biological merit is the main criterion for evaluation and comparison between the various biclustering methods.

In this chapter we attempt to develope a comparative tool (Bicat-Plus) which is showen in Figure 3 that includes the biological comparative methodology and to be as an extension to the BicAT program[26].

The Goal of BicAT-Plus is to enable researchers and biologists to compare between the different biclustering methods based on set of biological merits and draw conclusion on the biological meaning of the results. In addition, BicAT-Plus help researchers in comparing and evaluating the algorithms results multiple times according to the user selected parameter values as well as the required biological perspective on various datasets.

BicAT-Plus has many features, which could be summarized in the following:-

Algorithms required to be compared could be selected from the biclustering list (left list) to the compared list (right list). External biclustering results for other algorithms could be included in the comparison process. In addition, the organism model, selectable significance level, and GO category should be selected. Finally, Comparison criteria have to be selected based on the user biological metric.

1. User could perform biclusters functional analysis using the three Gene Ontology (GO) categories (biological process, molecular function and cellular component) (Figure3 with label number 1).
2. User could evaluate the quality of each biclustering algorithm results after applying the GO functional analysis and display the percentage of the enriched biclusters at different P-values (Figure3 with label number 2).
3. User could compare between the different biclustering algorithms according to the percentage of the functionally enriched biclusters at the required significance levels, the selected GO category and with certain filtration criteria for the GO terms. (Figure3 with label number 3).
4. User could evaluate and compare the results of external biclustering algorithms. This gives the BicAT-plus the advantage to be a generic tool that does not depend on the employed methods only. For example, it can be used to evaluate the quality of the new

algorithms introduced to the field and compare against the existing ones. (Figure 3 with label number 4).

5. User could display the results using graphical and statistical charts visualizations in multiple modes (2D and 3D).

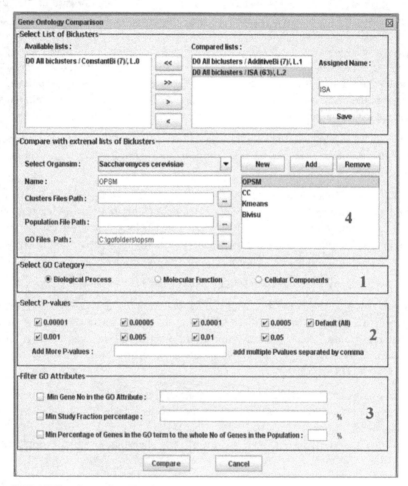

Figure 3. BicAT-Plus Comparison Panel.

2. Materials and methods

Before using the BicAT-Plus, Active Perl version 5.10 and Java Runtime Environment (JRE) version 6 are required to be installed on your machine. BicAT-Plus has been tested and show good performance on a PC machine with the following configurations: CPU: Pentium 4, 1.5 GHZ, RAM: 2.0 GB, Platform: windows XP professional with SP2.

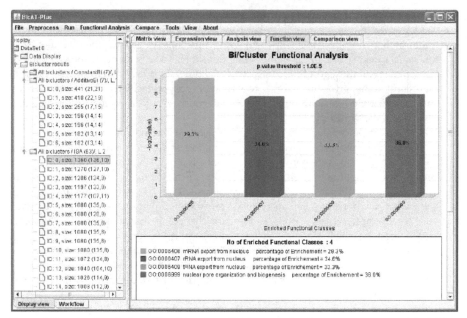

Figure 4. Functional analysis results of the selected bicluster. Each column represents an enriched GO functional class. And the height of the column is proportional to the significance of this enrichment (i.e. height = -log (p-value).

2.1. GO overrepresentation programs

Many programs like: BINGO[27], FUNCAT[28], GeneMerge[29] and FuncAssociate[30] were used to investigate whether the set of genes discovered by biclustering methods present significant enrichment with respect to a specific GO annotation provided by Gene Ontology Consortium [31]. BicAT-Plus used GeneMerge program as the most popular GO program. GeneMerge provides a statistical test for assessing the enrichment of each GO term in the sample test. The basic question answered by this test is as described by Steven *et al.*[27] "when sampling X genes (test set) out of N genes (reference set, either a graph or an annotation), what is the probability that x or more of these genes belong to a functional category C shared by n of the N genes in the reference set? The hypergeometric test, in which sampling occurs without replacement, answers this question in the form of P-value. It's counterpart with replacement, the binomial test, which provides only an approximate P-value, but requires less calculation time."

2.2. Comparative methodologies

BicAT-Plus provides reasonable methods for comparing the results of different biclustering algorithms by:

- Identifying the percentage of enriched or overrepresented biclusters with one or more GO term per multiple significance level for each algorithm. A bicluster is said to be significantly overrepresented (enriched) with a functional category if the P-value of this functional category is lower than the preset threshold. The results are displayed using a histogram for all the algorithms compared at the different preset significance levels, and the algorithm that gives the highest proportion of enriched biclusters for all significance levels is considered the optimum because it effectively groups the genes sharing similar functions in the same bicluster.
- Identifying the percentage of annotated genes per each enriched bicluster.
- Estimating the predictive power of algorithms to recover interesting patterns. Genes whose transcription is responsive to a variety of stresses have been implicated in a general Yeast response to stress (awkward). Other gene expression responses appear to be specific to particular environmental conditions. BicAT-Plus compares biclustering methods on the basis of their capacity to recover known patterns in experimental data sets. For example, Gasch et al.[32] measure changes in transcript levels over time responding to a panel of environmental changes, so it was expected to find biclusters enriched with one of response to stress (GO:0006950), Gene Ontology categories such as response to heat (GO:0009408), response to cold (GO:0009409) and response to glucose starvation(GO:0042149). The details of this comparison strategy are described in the results and in Table 3.

2.3. Comparison Process Steps

The following process diagram shown in Fig 5 summarizes the required steps by the user to compare between the different algorithms using the BicAT-plus:

1. Download BicAT-Plus from (*www.bioinformatics.org/bicat-plus/*).
2. Load Gene Expression Data to BicAT-Plus then run the selected five prominent biclustering methods with setting parameters as shown in Table 2.
3. Run GO comparison tool in the BicAT-Plus and add the available biclustering algorithms to the compared list as shown in Fig 1.
4. Select the available GO category e.g. biological process, molecular function and cellular components.
5. Select the P-values e.g. 0.00001, 0.0001, 0.01, 0.005, and 0.05.
6. Press compare button.
7. Press comparison menu, Functional enrichment and select 2D or 3D charts.

Bi/clustering Algorithm	Parameter settings
ISA	$t_g = 2.0$, $t_c = 2.0$, seeds = 500
CC	$\delta = 0.5$, $\alpha = 1.2$, $M = 100$
OPSM	$l = 100$
BiVisu	$E = 0.82$, $N_r = 10$, $N_c = 5$, $P_o = 25$
K-means	K=100

Table 2. Default Parameter settings of the compared bi/clustering methods. The definitions of these parameters are listed in their original publications [9, 15, 17-18, 20] respectively.

3. Results & discussion

The above comparison steps is performed on the gene expression data of *S. cerevisiae* provided by Gasch [32]. The dataset contains 2993 genes and 173 conditions of diverse environmental transitions such as temperature shocks, amino acid starvation, and nitrogen source depletion. This dataset is freely available from Stanford University website [33]. For each biclustering algorithm, we used the default parameters as authors recommend in their corresponding publications. See Table 2.

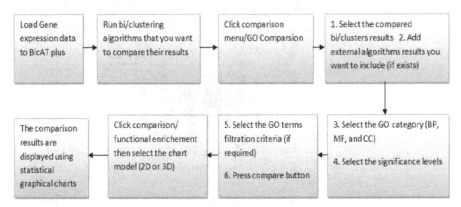

Figure 5. BicAT-Plus Comparison process steps

3.1. The percentage of enriched function

After applying the above steps on Gasch data[32] , BicAT-plus produce the histogram shown in Fig 6. Investigating this figure, we observed that OPSM algorithm gave a high portion of functionally enriched biclusters at all significance levels (from 85% to 100 %). Next to OPSM, ISA show relatively high portions of enriched biclusters.

In order to evaluate the ability of the algorithms to group the maximum number of genes whose expression patterns are similar and sharing the same GO category, we use the filtration criteria developed in the comparative tool by neglecting those bi/clusters which have study fraction less than 25%. The study fraction of a GO term is the fraction of genes in the study set (bicluster) with this term.

$$Study\ fraction\ of\ a\ GO\ term = \frac{No\ of\ genes\ sharing\ the\ GO\ term\ in\ a\ bicluster}{total\ number\ of\ genes\ in\ this\ bicluster} \times 100$$

Figure 7 shows that OPSM and ISA have highly enriched biclusters/clusters that have large number of genes per each GO category. On the other hand, Bivisu biclusters are strongly affected by this filtration and they contain a lower number of genes per each category. This filtration will help in identifying the powerful and most reliable algorithms which are able to group maximum numbers of genes sharing same functions in one bicluster.

3.2. The predictability power to recover interested pattern

The user could compare bi/clusters algorithms based on which of them could recover defined pattern like which one of them could recover bi/clusters which have response to the conditions applied in Gasch experiments. In Table 2, the difference between the biclusters/clusters contents were summarized.

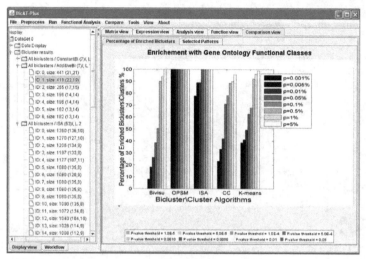

Figure 6. Percentage of biclusters significantly enriched by GO Biological Process category (*S. cerevisiae*) for the five selected biclustering methods and K-means at different significance levels p.

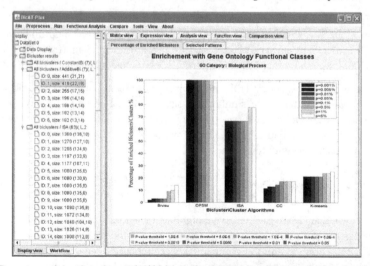

Figure 7. Percentage of significantly enriched biclusters by GO Biological Process category by setting the allowed minimum number of genes per each GO category to 10 and the study fraction to large than 50%.

Although OPSM show high percentage level of enriched biclusters (as shown in Fig 6, 7), its biclusters do not contain any genes within any GO category response to Gasch experiments. The k-means and Bivisu cluster/bicluster results distinguished a unique GO category, which is GO: 0000304 (response to singlet oxygen), and GO: 0042542 (response to hydrogen peroxide) The powerful usage of these bicluster algorithms is significantly appeared in GO: 0006995 "cellular response to nitrogen starvation" where these algorithms were able to discover 4 out of 5 annotated genes without any prior biological information or on desk experiments.

GO Term / (number of annotated genes)	K-means	CC	ISA	Bivisu	OPSM
GO:0042493 Response to drug / **(118)**	4	5	7	6	0
GO:0006970 response to osmotic stress / **(83)**	3	5	6	3	0
GO:0006979 response to oxidative stress / (79)	2	7	11	0	0
GO:0046686 response to cadmium ion / (102)	2	3	2	2	0
GO:0043330 response to exogenous dsRNA / (7)	2	3	2	2	0
GO:0046685 response to arsenic / (77)	2	0	2	2	0
GO:0006950 response to stress / (532)	9	11	16	2	0
GO:0009408 response to heat / (24)	3	0	2	2	0
GO:0009409 response to cold / (7)	0	0	2	0	0
GO:0009267 cellular response to starvation / (44)	0	2	0	0	0
GO:0006995 cellular response to nitrogen starvation / (5)	4	4	4	0	0
GO:0042149 cellular response to glucose starvation / (5)	0	2	0	0	0
GO:0009651 response to salt stress / (15)	2	7	0	0	0
GO:0042542 response to hydrogen peroxide /(5)	0	0	0	2	0
GO:0006974 response to DNA damage stimulus / (240)	0	22	0	3	0
GO:0000304 response to singlet oxygen / (4)	2	0	0	0	0

Table 3. Gene Ontology category per number of annotated genes of the Bicluster/cluster algorithm results for the experimental condition on Gasch Experiments[32].

4. Conclusion

We have introduced the BicAT-Plus with reasonable comparative methodology based on the Gene Ontology. To the best of our knowledge such an automatic comparison tool of the various biclustering algorithms has not been available in the literature. BicAT-Plus is an open source tool written in java swing and it has a well structured design that can be extended easily to employ more comparative methodologies that help biologists to extract the best results of each algorithm and interpret these results to useful biological meaning.

In other words, the algorithms that show good quality of results (per the dataset) can be used to provide a simple means of gaining leads to the functions of many genes for which information is not available currently (unannotated genes).

Using BicAT-Plus, we can identify the highly enriched biclusters of the whole compared algorithms. This might be quite helpful in solving the dimensionality reduction problem of the Gene Regulatory Network construction from the gene expression data. This problem originates from the relatively few time points (conditions or samples) with respect to the large number of genes in the microarray dataset.

Finally there are several aspects of this research that worth further investigation, according to the Studies carried out so far and also introducing new ideas for consideration

1. Enrich the BicAT-Plus with more comparative methodologies beside GO. For example, KEGG and promoter analysis by identifying the transcription factors for the clustered genes.
2. Extend the BicAT-Plus to provide users with multiple export options for the interested enriched biclusters.
3. Embed the BicAT-Plus as a plug-in in the Cytoscape platform[34] which is open source bioinformatics software for visualizing molecular interaction networks and biological pathways and integrating these networks with annotations, gene expression profiles and other state data. Thus, very promising challenge is to get use of the highly enriched biclusters identified by the BicAT-Plus in solving these integrated networks in the Cytoscape.

Author details

Fadhl M. Al-Akwaa
Biomedical Eng. Dept., Univ. of Science & Technology, Sana'a, Yemen

5. References

[1] Fung G: A Comprehensive Overview of Basic Clustering Algorithms. *Citeseer* 2001:1-37.
[2] Sharan R, Elkon R, Shamir R: Cluster analysis and its applications to gene expression data. *Ernst Schering Res Found Workshop* 2002:83-108.

[3] Eisen MB, Spellman PT, Brown PO, Botstein D: Cluster analysis and display of genome-wide expression patterns. *Proceedings of the National Academy of Sciences of the United States of America* 1998, 95:14863 - 14868.

[4] P. Tamayo DS, J. Mesirov, Q. Zhu, S. Kitareewan, E. Dmitrovsky, E. S. Lander, and T. R. Golub: Interpreting patterns of gene expression with self-organizing maps: Methods and application to hematopoietic dierentiation. In: *Proceedings of the National Academy of Sciences of the United States of America,: 1999*. 2907–2912.

[5] Sharan RSaR: Click: a clustering algorithm for gene Expression analysis. In: *Proceedings of the 8th International Conference on Intelligent Systems for Molecular Biology: 2000*. 307–316.

[6] Tavazoie S, Hughes JD, Campbell MJ, Cho RJ, Church GM: Systematic determination of genetic network architecture. *Nature Genetics* 1999, 22:281-285.

[7] Johnson S: Hierarchical clustering schemes. *Psychometrika* 1967, 32(3):241-254.

[8] Madeira SC, Oliveira AL: Biclustering algorithms for biological data analysis: a survey. *IEEE/ACM Trans Comput Biol Bioinform* 2004, 1(1):24 - 45.

[9] Cheng KO, Law NF, Siu WC, Lau TH: BiVisu: software tool for bicluster detection and visualization. *Bioinformatics* 2007, 23(17):2342 - 2344.

[10] Liu X, Wang L: Computing the maximum similarity bi-clusters of gene expression data. *Bioinformatics* 2007, 23(1):50-56.

[11] Prelic A, Bleuler S, Zimmermann P, Wille A, Buhlmann P, Gruissem W, Hennig L, Thiele L, Zitzler E: A Systematic comparison and evaluation of biclustering methods for gene expression data. *Bioinformatics* 2006, 22(9):1122 - 1129.

[12] A. Tchagang and A. Twefik: Robust biclustering algorithm (ROBA) for DNA microarray data analysis. In: *IEEE/SP 13thWorkshop on Statistical Signal Processing*. 2005: 984–989.

[13] Murali TM, S K: Extracting conserved gene expression motifs from gene expression data. In: *Pac Symp Biocomput*. 2003: 77–88.

[14] A. Tanay RS, M. Kupiec, and R. Shamir, : Revealing modularity and organization in the yeast molecular network by integrated analysis of highly heterogeneous genomewide data,. In: *Proceedings of the National Academy of Sciences of the United States of America: 2004*. 2981–2986.

[15] Ben-Dor A, Chor B, Karp R, Yakhini Z: Discovering local structure in gene expression data: the order-preserving submatrix problem. *Journal of Computational Biology* 2003, 10:373 - 384.

[16] H. Wang WW, J. Yang, and P. S. Yu, : Clustering by Pattern Similarity: the pCluster Algorithm. *SIGMOD* 2002.

[17] Ihmels J, Friedlander G, Bergmann S, Sarig O, Ziv Y, Barkai N: Revealing modular organization in the yeast transcriptional network. *Nature Genetics* 2002, 31:370 - 377.

[18] Cheng Y, Church GM: Biclustering of expression data. *Proceedings of 8th International Conference on Intelligent Systems for Molecular Biology* 2000:93 - 103.

[19] Hartigan J: Direct Clustering of a data matrix. *Journal of the American Statistical Association* 1972, 67:123–129.

[20] Ihmels J, Bergmann S, Barkai N: Defining transcription modules using large-scale gene expression data. *Bioinformatics* 2004, 20:1993 - 2003.

[21] Tavazoie S, Hughes J, Campbell M, Cho R, Church G: Systematic determination of genetic network architecture. *Nature Genetics* 1999, 22:281-285.

[22] Guthke R, Moller U, Hoffmann M, Thies F, Topfer S: Dynamic network reconstruction from gene expression data applied to immune response during bacterial infection. *Bioinformatics* 2005, 21(8):1626-1634.

[23] D'haeseleer P, Liang S, Somogyi R: Genetic network inference: from co-expression clustering to reverse engineering. *Bioinformatics* 2000, 16(8):707-726.

[24] Reiss D, Baliga N, Bonneau R: Integrated biclustering of heterogeneous genome-wide datasets for the inference of global regulatory networks. *BMC Bioinformatics* 2006, 7(1):280.

[25] Yip KYaQ, Peishen and Schultz, Martin and Cheung, David W and Cheung, Kei-Hoi: SemBiosphere: A Semantic Web Approach to Recommending Microarray Clustering Services. In: *The Pacific Symposium on Biocomputing.* 2006: 188-199.

[26] Barkow S, Bleuler S, Prelic A, Zimmermann P, Zitzler E: BicAT: a biclustering analysis toolbox. *Bioinformatics* 2006, 22(10):1282-1283.

[27] Maere S, Heymans K, Kuiper M: BiNGO: a Cytoscape plugin to assess overrepresentation of Gene Ontology categories in Biological Networks. *Bioinformatics* 2005, 21(16):3448-3449.

[28] Ruepp A, Zollner A, Maier D, Albermann K, Hani J, Mokrejs M, Tetko I, Guldener U, Mannhaupt G, Munsterkotter M *et al*: The FunCat, a functional annotation scheme for systematic classification of proteins from whole genomes. *Nucl Acids Res* 2004, 32(18):5539-5545.

[29] Castillo-Davis CI, Hartl DL: GeneMerge - post-genomic analysis, data mining, and hypothesis testing. *Bioinformatics* 2003, 19(7):891 - 892.

[30] Berriz GF, King OD, Bryant B, Sander C, Roth FP: Characterizing gene sets with FuncAssociate. *Bioinformatics* 2003, 19(18):2502-2504.

[31] Ashburner M, Ball C, Blake J, Botstein D, Butler H, Cherry J, Davis A, Dolinski K, Dwight S, Eppig J *et al*: Gene ontology: tool for the unification of biology. The Gene Ontology Consortium. *Nat Genet* 2000, 25:25 - 29.

[32] Gasch AP, Spellman PT, Kao CM, Carmel-Harel O, Eisen MB, Storz G, Botstein D, Brown PO: Genomic Expression Programs in the Response of Yeast Cells to Environmental Changes. *Mol Biol Cell* 2000, 11(12):4241-4257.

[33] http://genome-www.stanford.edu/yeast/_stress

[34] Shannon P, Markiel A, Ozier O, Baliga N, Wang J, Ramage D, Amin N, Schwikowski B, Ideker T: Cytoscape: a software environment for integrated models of biomolecular interaction networks. *Genome Res* 2003, 13(11):2498-2504.

Chapter 3

Beyond the Gene List: Exploring Transcriptomics Data in Search for Gene Function, Trait Mechanisms and Genetic Architecture

Bregje Wertheim

Additional information is available at the end of the chapter

1. Introduction

Since the start of genomics research, genome-wide expression studies have been used prolifically as a tool to improve our understanding of the involvement of genes in various biological processes. Measuring gene expression patterns simultaneously across all the genes in the genome, i.e. transcriptomics, is a uniquely powerful technology to explore potential novel candidate genes for a particular process. This genome-wide approach has the huge advantage that we do not have to specify in advance which genes we believe to be involved, and as such, we are not limited by our current knowledge. Transcriptomics is an important first step to study traits that are under the control of several to many genes (i.e., polygenic traits) and responsive to external conditions and internal states (i.e., multifactorial traits).

The identification of potential novel candidate genes, however, is only a limited part of the power of transcriptomics. With this technology, the expression of thousands of genes is measured simultaneously. It provides a snapshot of all genes that are actively transcribed during a particular process. When we compare these measurements between conditions or treatments, those genes that are expressed at higher or lower level under a particular condition can be identified. As such, transcriptomics maximizes the awareness of effects anywhere in the genome, including those associated by costs, trade-offs and epistatic interactions. This could be viewed as a complication of transcriptomics data, because a change in expression does not necessarily reflect a causal relationship to the process of interest. In fact, however, it is also one of the major strengths of this technology. By combining various bio-informatic tools and resources, it is possible to obtain an insight into intricate gene-interaction networks, the regulatory control of traits, and the implications of a trait or process on the full phenotype.

In functional genomics, transcriptomics studies are typically a comparison between biological samples (e.g., a cell type, organ, individual, or group of individuals) that were collected under different conditions, to analyse which genes were up-regulated or down-regulated (i.e., were expressed at higher, respectively, lower levels) in response or relation to the condition. These conditions can be experimentally induced (e.g., treatment *versus* control, different dosages of a chemical, different food conditions or temperatures, etc.), or they represent different natural stages (e.g., diseased *versus* healthy, male *versus* female, different developmental stages or aged individuals, different genotypes, different tissues, different epigenetic profiles, etc). Including a proper control treatment or reference is crucially important for the interpretation of gene expression differences that results from such a comparison. There will always be a large number of genes expressed in any biological sample, and without control or reference, it is impossible to attribute expression of particular genes to the condition of interest. The purpose of transcriptomics is to reveal how the expression patterns *change* under different conditions.

Transcriptomics technology is used to characterize the composition of the messenger RNA (mRNA) pools from each biological sample. The mRNAs are the transcripts of a gene that carry the information encoded in the gene to the site of protein synthesis. When a particular mRNA is present in a biological sample, it implies that the corresponding gene was expressed, and a template is available for the synthesis of the protein product of that gene. The abundance of each mRNA in the pool represents the level of expression of the corresponding gene. By comparing the relative proportional representation of each mRNA in the total mRNA pool among the samples, we can identify which genes differed in expression in response or relation to the compared conditions. The most widely use technological platform for whole-genome expression studies are microarrays, although the sequencing of the transcriptome is rapidly increasing in popularity (Figure 1).

Microarrays are solid-based platforms (e.g., glass slides), containing millions of copies for thousands of 'reporter probes' that comprise part of the sequences of the genes in the genome. By binding (or 'hybridizing') fluorescent-labelled copies of the original mRNAs to the probes, measuring the label intensities for each position on the array, and associating these positions to their specific reporter probes, one can infer the presence and abundance of each transcript in the labelled RNA pool (Figure 1). It is assumed this representation is proportional to their abundances in the original mRNA samples. Microarrays are relatively cheap, and the tools to analyse the data have been developed, matured and tested. This makes microarrays an affordable and accessible platform for many applications [1]. After the initial introduction of expression arrays that reported only on known or predicted genes, tiling arrays were developed that contained reporter probes across the full genome, including the non-coding, non-translated and non-transcribed chromosomal regions. This enabled the identification of novel transcripts, including non-coding RNA genes, as well as a better characterization of splice variants and exons [2].

The latest developments in next-generation sequencing technologies are making transcriptome sequencing more affordable, and they provide a number of advantages over microarrays [3]. For this approach, the mRNA pool is converted into cDNA (either wholly or after a partial digestion), which is then used as template for high-throughput sequencing. The generated sequence information is mapped to, or assembled into, a reference transcriptome, and the number of sequence copies generated for each gene is used to infer the number of mRNA copies in the original sample (Figure 1). Sequencing approaches provide more comprehensive information on the transcript characteristics (e.g. splice variants, mRNA sequence variations, gene fusions, etc.), they are not limited to the known or predicted genes of an organism or the genes represented on an microarray, and they avoid some problems inherent to slide-based technologies [4]. A downside of transcriptome sequencing is that the Quality Control and pre-processing and analysis procedures for these data have not yet fully matured, and the assembly of, or mapping against, the reference transcriptome requires substantial computing power, making this technology still less accessible.

In essence, both technological platforms yield data of very similar nature, although the information of sequencing approaches may be more specific and detailed than array-based approaches. After the specific pre-processing that each platform requires, the data can be analysed with similar methods, leading to a list or ranking of genes that show changes in expression patterns or transcript characteristics (e.g. splice variants) among the compared conditions. As such, the gene list provides a first step to identify the genes that potentially matter or are affected by a particular condition. A change in expression, however, is insufficient evidence for establishing a clear link between a gene and the trait of interest. At best, the genes on the list may be associated with the trait or condition of interest, while causality or direct involvement in the trait still needs to be established through additional empirical approaches.

Before discussing how gene lists can be generated or used for further analysis, it is important to emphasize that certain limitations are inherent to transcriptomics data. These limitations can be specific to the used platform, for instance microarrays can only report on the activity of genes that are known or represented on the array. Most limitations, however, are irrespective of the technology. As mentioned, genes that are differentially expressed are not necessarily causal to a particular trait or response. Moreover, not all the genes that are involved in a response or trait are detected by a changes in expression. Any post-transcriptional modifications or non-transcriptional processes (such as the re-directing of a transcription factor from its regular processes towards another function) are typically not detectable by a change in gene expression. A further precautionary note is warranted for the design, set-up and execution of any transcriptomics study. An essential requirement for associating changes in gene expression among different samples to a particular condition or treatment of interest, is to ensure that the only difference is the condition or treatment of interest. For example, the collection of control and treatment samples should be done simultaneously (e.g., not before infection and 12 hours after infection) by the same person, to avoid that circadian rhythms or handling effects differ between the samples. When such precautions would not be taken, genes responsive to the treatment would be confounded with genes responsive to these extraneous factors. It is impossible to resolve such confounding effects after the measuring of gene expression. The only way to avoid such

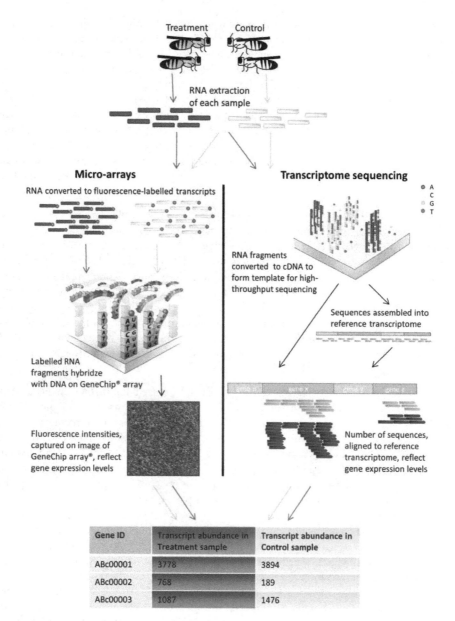

Figure 1. Schematic overview of transcriptomics approaches, using microarrays or transcriptome sequencing. Although the technologies differ, both approaches compare all the mRNAs in biological samples under different conditions, and provide quantifications of the abundance of all gene transcripts for each sample. *Images of GeneChips® courtesy of Affymetrix.*

issues is to take due care during experimental design, sample collection and sample preparation. Despite these limitations that are inherent to any transcriptomics technology, the resulting data does provide an array of possibilities for further meaningful analysis.

In this chapter, I will illustrate various ways in which transcriptomics data can be analysed, to identify novel candidate genes for the process of interest, and additionally, how to move beyond this list of candidate genes towards the molecular mechanisms and gene interaction networks of a trait. For these illustrations, I will mostly use transcriptomics data on the innate immunity in *Drosophila* larvae after parasitism. Our analysis on the transcriptomics during the acute immune response to infection by parasitic wasps [5], as well as between strains that differ genetically in resistance to these parasites [6], revealed a complex gene interaction network associated with defense mechanisms. The immune response to parasites is triggered by recognizing the invasion of the parasite, and comprises of the proliferation and differentiation of specialized blood cells that surround the parasite in a multi-layered capsule, and sealing the capsule with a layer of melanin. This melanotic encapsulation sequesters and kills the parasite [7]. By integrating the data from our studies with various resources and bioinformatics approaches, we gained a more comprehensive insight in the interactive and regulatory network of genes that are associated with the immune response to parasitism. We identified shared regulatory elements among genes that showed similar expression patterns, physiological costs associated with evoking the immune response, chromosomal positions that were associated with resistance traits and indications for epistatic gene-interactions. Combined, this information provided us with new insights on the mechanisms and complex genetic architecture of the innate immune response.

2. Constructing a list of genes with differential expression

The fundamental purpose of a transcriptomics experiment is to identify the genes with changed expression under a particular condition, which is done by comparing the abundance measurements for each gene transcript among the biological samples. Depending on the platform used, these abundance measurements are derived from fluorescence intensity measurements captured in digital images of the microarrays, or the counts of the number of transcripts for sequencing approaches (Figure 1). These measurements, however, are not only reflecting the biologically interesting variation in gene expression under the different conditions, but also a substantial level of technical variation that is introduced during the preparation and measuring of the samples. This includes, for example, residues of reagents that create a background signal on microarrays, short fragments of RNA that bind non-specifically to microarray probes or cannot be uniquely mapped to a reference genome, slight differences in RNA doses for the different samples, or slight differences among samples/batches in the efficiency of the molecular techniques. Some of these aspects affect whole samples, while others are specific to particular genes. To perform the meaningful comparisons on the variation in gene expression measurements, it is typically essential to first eliminate the bias introduced by technical variation as much as possible.

The raw intensity measurements first need to be pre-processed to deal with the technical variation, normalized to scale all samples to the same range, and combined into a single expression value per gene per sample for comparisons. Many different approaches have been developed for the pre-processing and normalization of microarrays, and subsequent studies have tried to determine the optimal strategy to remove the noise without introducing bias. Some approaches are outperforming others and consensus has been mostly reached for the commonly used microarray platforms, although full consensus for all microarray platforms is still lacking [8]. Also for transcriptome sequencing, normalization is important to address deviations due to slight differences in doses, the gene length and GC-content. The exploration of the best pre-processing and normalization approaches for transcriptome sequencing are still being established (e.g., [4,9]).

To statistically test for changes in gene expression, biological replication is essential. Having multiple biological units for each condition enables the estimation of variation within and between conditions, which allows for the partitioning of all variation into noise (i.e., technical and random variation), and the biologically interesting variation reflecting the changes in gene expression patterns. Technical replications are sometimes also incorporated in the platform or analysis, for example by repeating the same probes on a microarray, by applying a dye-swap on samples, or by testing the same samples twice. Although this can increase the accuracy and sensitivity for the estimation of technical variation, it is generally not as important as biological replication is for increasing the sensitivity and power of the analysis. The minimum number of replications that is required for any transcriptomics experiment depends, among others, on the objective of the experiment, the required sensitivity, the type of microarray or sequencing method used, the experimental design, and the number of treatment groups [10]. Measuring gene expression across a time course may also be a powerful way to increase the power of the analysis, as well as providing a means to determine the sequence of action for genes.

For the statistical analysis of transcriptomics data, many different alternatives are available. Most tests developed for microarray data or transcriptome sequencing are essentially modifications of more standard statistical tests [8]. To identify the genes showing differential expression (i.e., differences in expression level) among treatments or conditions, many of the statistical procedures consist of some form of variance analysis and test whether the variance in expression patterns among treatments or conditions exceeds the variance between biological replicates within a treatment. The most commonly used tests include (modifications of) t-tests, ANOVAs, regression analysis, mixed models and generalized linear models. The modifications for these tests are primarily to increase power for the often small sample sizes, and to avoid violation of the assumptions for the parametric tests, in particular the assumptions of a Normal distribution and independence among measurements. Modifications include methods to shrink variance estimates (using combined information on variance for the large number of measurement on a single sample), permutation approaches and empirical Bayesian methods. Similar to the best choice for the number of biological replicates, the best statistical approach depends on the objective of the experiment, the transcriptomics platform used, the experimental design, the number of treatment groups and the number of replicates per treatment.

Not only statistical significance, but also the magnitude of a change in expression (or the 'fold change') between conditions is often provided, sometimes as an auxiliary for biological significance. Fold changes are typically provided at a log2 scale, so that the fold changes are centred around zero, and a doubling or halving of expression level in the treatment compared to the control would result in an equal deviation from zero. These fold change data can be plotted to visualize the differentially expressed genes, either in relation to the average expression level of that gene (MA-plot, Figure 2a), or in relation to the statistical significance (volcano plot, Figure 2b). It should be realized, however, that fold changes are fickle indicators of biological significance. Firstly, depending on the position and role of a particular gene in a regulatory network (e.g., a central transcription factor, or a direct regulator of transcriptional activity), a small fold change may have large biological implications. Large fold changes could be primarily expected at the margins of these networks, which may involve the final effectors of the response while that may reveal little about the key regulators of the response. Secondly, microarrays typically only detect large fold changes in the intermediate range of expression values. Low levels of expression may be below the detection limit of the array, and background noise or corrections may obscure any changes in the expression of such genes. High levels of expression may result in saturation of the probes, vastly underestimating the actual fold changes. Transcriptome sequencing approaches would not be biased towards these intermediate expression levels, but instead, could suffer from exaggerated fold-change estimates for genes not expressed, or expressed at very low level, in one sample or both samples (when the denominator approaches zero).

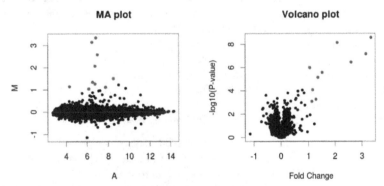

Figure 2. Plots that summarize the fold-change differences in gene expression between two conditions. a) MA plots portray for each gene the average gene expression across the two conditions on the x-axis (A), and the log2 fold change difference in expression between the two conditions on the y-axis (M). b) Volcano plots portray for each gene the log2 fold change in difference of expression between the two conditions on the x-axis (Fold Change) and the statistical significance for the *t*-test on expression measurements between the two conditions on the y-axis (–log10 P-value). The presented data is on *Drosophila* larvae 12 hours after being parasitized and control larvae (that had not been parasitized) [5]. The 'outliers' in both plots represent genes that differed in expression between the two conditions. In red are the genes that both scored a P-value < 0.001 and had at least a 2-fold change in expression between the two conditions. Applying these combined criteria for assigning significance would exclude several 'outlier' genes with high average expression levels (a) and/or with low p-values (b).

Finally, to determine the genes with significant differences in expression among conditions or treatments, a statistical correction needs to be applied for the large number of statistical tests for each experiment (i.e., multiplicity or multiple testing). In a transcriptomics experiments, several thousands of genes are tested, and each gene is analysed for differences in expression among conditions. In statistics, we normally use a type I error rate of $\alpha = 0.05$, which means that we accept that in 5% of cases where we rejected the null hypotheses (H_0: no differences among conditions) and called something 'significantly' different, the observed difference was purely by chance. When we do not correct the type I error rate while performing thousands of statistical tests (i.e., one for each gene), this would result in hundreds of genes called significantly differentially expressed, while these differences were merely by chance. Genes that are deemed differentially expressed while they are not, are *false positives*. Genes that are deemed not differentially expressed while they are, are *false negatives*. Correcting for false positives in large scale experiments is needed to avoid including many erroneous calls, but it needs to be carefully balanced by controlling for false negatives to ensure optimal sensitivity and accuracy of the analysis.

The typical statistical correction for false positives in non-genomic experiments with multiple testing is a Bonferroni correction, which divides α by the number of statistical tests applied to the data. This approach, however, is often too conservative (i.e., accepting the null hypothesis H_0, while it was false) for the thousands of tests in transcriptomics analyses, and would result in a large number of false negatives. The most widely used correction for multiple testing in transcriptomics analysis is a False Discovery Rate (FDR) correction, which attempts to provide a more even balance between false positives and false negatives. Several FDR approaches exist, but they generally adjust or replace the P-value for significance to reflect the likely proportion of false positives among the genes that are called significant. For example, when we would identify 100 genes with an FDR adjusted P-value (P_{adj} or q-value) of <0.05, we would on average expect less than 5 of these genes to be false positives [11]. The acceptance level for significance used with FDR often ranges from P_{adj} <0.001 to <0.10, depending on the desired sensitivity and accuracy, the sample size (i.e., power) and the estimated numbers of genes with differential expression.

The end result of all pre-processing steps, normalisation, statistical analyses and corrections for false positives is a list or ranking of genes that significantly changed expression in response or relation to the different conditions that were compared. This lists contains potential candidate genes that may be actively involved in the process of interest. However, many genes are also included in the list that are only indirectly associated with the response or process of interest. Moreover, the gene list does not contain *all* the (candidate) genes that are involved in the process, but only these that could be detected by transcriptomics and under the particular experimental conditions (e.g. time points during the response, sample sizes, technological platform) and analysis choices (e.g. normalization approach, acceptance thresholds for significance). Finding gene expression changes in a transcriptomics experiments is not required, nor sufficient, evidence for the function of a gene or its involvement in a biological process. It is, however, a valuable starting point for further analysis.

3. Standard explorations of the gene list

The first inspection of a gene list typically is to link the gene names to what is known, predicted and published about these genes, both in terms of the function of the gene (product), the protein family or protein domains that the gene codes for, and the signal transduction pathways in which it participates. For model species and other species for which the full genomic sequence is available, repositories exist that combine several sources of information on individual genes (for example, see "www.nature.com/scitable/content/ Genomics-Databases-744357" for a list of species-specific repositories [12]). The annotation of genes is mostly following a controlled vocabulary or restricted terminology. For functional annotations, Gene Ontology (GO) is a widely used vocabulary. Gene Ontology describes the genes and their products (e.g., the proteins for which a gene codes) within three main Ontology domains: Molecular Function, Biological Process and Cellular Component. Genes can be described at various hierarchical levels using this GO terminology, ranging from broad over-arching themes to very specific descriptions. Descriptions of protein domains are often inferred based on sequence similarity to other organism, for example using the InterPro terminology. Since many proteins are involved in several biological processes or contain more than one functional domain, genes (or gene products) have often different GO annotations across the three GO domains and different IP annotations (Table 1).

Gene Name (symbol)	Gene Ontology Annotation	InterPro Annotation
αPS4	Cellular Component: Integrin complex Biological Process: Cell adhesion Biological Process: Cell-matrix adhesion Biological Process: Heterophilic cell-cell adhesion Molecular Function: Cell adhesion molecule binding Molecular Function: Receptor activity	Integrin alpha chain Integrin alpha beta-propellor Integrin alpha-2 Integrin alpha chain, C-terminal cytoplasmic region, conserved site FG-GAP
lectin-24A	Cellular Component: - Biological Process: Galactose binding Molecular Function: -	C-type lectin C-type lectin fold
Thiolester containing protein II (TepII)	Cellular Component: Extracellular space Biological Process: Antibacterial humoral response Biological Process: Defense response to gram-negative bacterium Biological Process: Phagocytosis, engulfment Molecular Function: Endopeptidase inhibitor activity Molecular Function: Peptidase inhibitor activity	Terpenoid cylases/protein prenyltransferase alpha-alpha toroid Alpha-macroglobulin, receptor-binding Alpha-2-macroglobulin, N-terminal Alpha-2-macroglobulin, N-terminal 2 A-macroglobulin complement component Alpha-2-macroglobulin, conserved site Alpha-2-macroglobulin, thiol-ester bond-forming

Table 1. Examples of gene annotations, using the vocabulary of the Gene Ontology (GO) and InterPro (IP). Annotations are provided for three genes that were differentially expressed during the immune response of *Drosophila* after infection by parasites [5]. The GO annotations describe the function and

process that have been reported for the protein, and the IP annotations describe the protein domains. Genes that are involved in different processes, or coding for proteins with multiple functional domains, may contain a variety of annotations. Many genes, however, are not fully annotated.

The abundance and reliability of annotation information is highly variable among genes and species: some genes are well studied and annotations are solidly supported by empirical evidence, while other genes are not annotated, only partially annotated or annotations are based only on unconfirmed computer predictions or non-traceable author statements. Furthermore, for model organisms the functional annotations have accumulated by the studies of many researchers over long periods, while for non-model organisms or new model organisms, there is often only limited detailed knowledge available. Yet, even for these non-model organisms, various resources exist that enable high level analysis of transcriptomics data based on homologies, such as, for example, the Blast2GO suite [13].

Gene lists from transcriptomics experiments are particularly amenable for enrichment analyses of functional annotations. An enrichment of a particular functional annotation implies that it is represented more often among the gene list members than would be expected by chance alone, based on the proportion of the genes in the genome with that annotation. Multiple interfaces and online tools have been developed for this purpose (e.g., DAVID for large gene lists [14] and Catmap for gene lists that are ranked for significance, but without actually applying a significance threshold cutoff [15]). When the conditions or treatments of interest resulted in a coordinated response in the gene interaction network, the likelihood increases of finding genes with changed expression sharing the same annotation. Such enrichments may be informative for identifying different biological processes or protein families that are associated with, or affected by, a response to the condition or treatment of interest. This may also be informative to identify possible costs or trade-offs that are associated with the response. For example, within the gene list for the response to parasite infection [5], we identified a set of genes involved in puparial adhesion. These genes were expressed at lower levels in the infected larvae at 72 hours after infection, and reflect the delay in development these larvae incurred by investing energy and resources in the immune response.

The list of differentially expressed genes can be compared to other gene lists, which could be derived from other transcriptomics studies, known candidate genes for the process of interest, or any other approach that identified a set of genes associated with a particular condition. Venn diagrams can summarize these gene list comparisons (Figure 3). Reporting how many of the genes were shared with other gene list(s), and how many are unique for each gene list, provides a quick overview of the numbers of genes that may be of particular interest. Sometimes it is the genes that are also present in the other gene list(s) that are of particular interest, for example when multiple sources of evidence are combined or to identify cross-talk between gene interaction networks. Alternatively, one could focus on the unique genes to identify novel candidate genes that had not previously been associated with the process of interest.

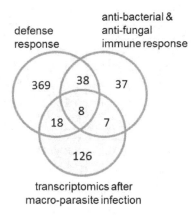

Figure 3. Venn diagram of differentially expressed genes in *Drosophila* larvae after infection by a parasitic wasp, and genes that have been previously implicated in defense responses and anti-microbial immune responses. Infection by a parasitic wasp ('macro-parasite') triggers a cellular immune response that is substantially different from general defense responses and the mostly humoral immune response against bacterial and fungal infections ('micro-parasites'). This is reflected both in the relatively large number of known immunity genes that did not change expression after infection with macro-parasites, and in the large number of differentially expressed genes after macro-parasite infection that had not previously been associated with immunity and defense. *Redrawn with permission after* [5], *first published by BioMed Central.*

When several conditions or time points are included in the experimental design, clustering the genes according to their expression pattern across these conditions or time points allows for identifying groups of genes that responded similarly, and analysing these separately from genes with different behaviour. An enrichment analyses on such groups of genes may identify a common theme to groups with a particular expression profile across the conditions or time points. For example, in our transcriptomics study after infection with macroparasites, we identified groups of genes with a peak in up-regulated expression 1-6 hours after infection, at 6-24 hours after infection, and at 24-72 hours after infection, and groups with down-regulated expression either throughout the time course, or at 72 hours after infection (Figure 4). The first group of genes was enriched for immunity genes (clusters 1 and 2), the second group of genes for proteolysis and serine-type endopeptidases (cluster 12), and the last group in puparial adhesion (cluster 9). These patterns can be used both to get a more detailed profile for the various processes that occur during the response. Additionally, it may serve as a starting point for inferring the functions of unannotated genes. For example, the *Drosophila* genome codes for 201 genes with serine-type endopeptidase activity, which function in development, immunity and various other biological processes. Only 22 of these genes had been functionally annotated with a role in immunity, but unannotated serine-type endopeptidase genes that responded similarly to infection as genes with a functional annotation in immunity or defense could be putatively assigned the same functions [16].

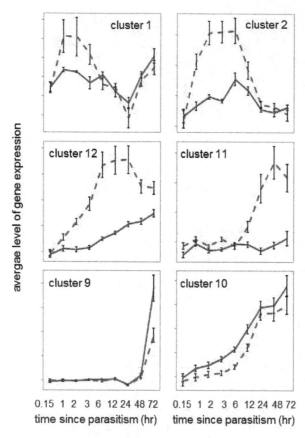

Figure 4. Clustering of genes that show similar expression patterns in *Drosophila* larvae across the 72 hour time course after infection by parasitic wasps. The average expression levels (± standard errors of the mean) for the genes within the clusters (log2 transformed and divided by the median expression level for that gene across all time points) is shown. Dashed red lines represent the gene expression in parasitized larvae and solid blue lines represent the gene expression in control (not parasitized) larvae. *Partially redrawn with permission after* [5] *, first published by BioMed Central.*

In addition to these general approaches for any transcriptomics analysis, regardless of the platform that was used, some additional insights could be gained from using tiling arrays or transcriptome sequencing. Not only the expression level could be determined for each gene, but also alternative isoforms of transcripts, including splice variants and sequence variations (either in the coding regions or in the untranslated regions of the transcripts). In humans, transcriptome sequencing revealed that splicing isoforms from various tissues showed systematic differences, including exon skipping, alternative 3' or 5' splice sites, mutually exclusive exons and alternative first or last exons [17]. New methods allow for the quantification of gene expression levels for the individual isoforms, which can improve the accuracy of expression measures and provide details on the role of the untranslated regions in gene expression regulation [18].

4. Beyond the gene list

The descriptions of the analyses so far have centred on querying repositories containing the functional annotations for genes, to explore what is known on the genes in the gene list and what additional light this may shed on sub-processes, the unannotated genes and associated responses. Yet, many additional resources and genomic databases are available that may be cross-referenced and combined with the gene list, to obtain additional information on these genes and their interactions. Rather than focussing on individual members of the gene list and what is known, these approaches search for emergent properties of the gene list. Especially when the organism that is studied is a model organism for which many sources of additional information are publicly available, there is a large array of possibilities for further analyses.

In addition to searching in specific repositories for functional annotations of genes, the extraction of information on genes and proteins from text documents (e.g., scientific papers) can leap across the boundaries of scientific disciplines. Text mining is the automated extraction of information on proteins or genes from a large literature collection (such as PubMed). It searches for associations between proteins and functional descriptors in the text. These descriptors can be of molecular origin to describe the annotations of the protein (as in the repositories), but also of a physiological, phenotypic or pathological origin to describe the inferences for the organism, or of phylogenomics origin related to the evolution of a gene. Through this additional dimension of information, text mining can help, for instance, to identify associations of the protein to rare mutations that are implicated in diseases, or to protein-protein interactions and regulatory pathways [19]. Text mining is different from a typical literature search, in that it not simply lists the hits, but parses the retrieved information according to further specifications (Figure 5). Various tools are available online (see for example www.ebi.ac.uk/Rebholz/resources.html for an overview).

Physiological responses or the focal tissue of a response to the treatment or condition of interest may also be explored through analysis of the gene list. For some model organisms, a tissue atlas is publicly available that specifies the level of expression for each gene in all tissues and/or developmental stages (e.g., FlyAtlas, Human Atlas Suite and eMouse Atlas). A large fraction of genes in the genome are not expressed homogeneously throughout the body, but show high specificity for particular tissues [21]. Using this information provides a means to screen for tissues that may contribute disproportionally to the response. For example, when the gene list is enriched for genes that are primarily expressed in a particular tissue (e.g. testes, brain, liver or salivary glands), this could indicate that these tissues are most severely affected or responding to the treatment of interest. Additionally, the atlases have raised an awareness for experimental design in transcriptomics studies: when the transcriptomics responses are localized in a particular (minor) tissue, it is difficult to detect

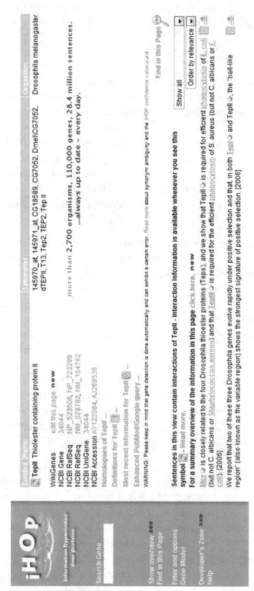

Figure 5. Example of the output from a text mining tool, iHOP [20], for one of the genes that was differentially expressed in *Drosophila* larvae after parasite infection. The functional annotations for the same gene, TepII, are summarized in Table 1. The text mining tool provided additional information on the evolution of the gene through information on related genes (paralogs) and domains of the gene that show signs of positive selection. *Screenshot retrieved from "iHOP - http://www.ihop-net.org/".*

accurate expression differences when the tissue is not studied in isolation. The chances of missing or underestimating the change in gene expression in mixed-tissue comparisons, or inappropriate tissues, are substantial.

To gain insight in the regulatory control of the response to the treatment or condition, a screen for *cis*-regulatory elements in the upstream regions of genes with differential expression may reveal transcription factors and/or co-factors that are involved. These *cis*-regulatory elements can consist of Transcription Factor Binding Motifs (TFBM), promotors, enhancers, silencers and other sequence motifs that regulate the genes [22]. To identify (putative) *cis*-regulatory elements, one could search for known sequence motifs (e.g., TFBMs or promotors) within a specified region upstream of the start codon and in the first intron. Several databases exist (for example, TRANSFAC, RegTransBase and JASPAR) that contain the published TFBMs and promotors. As the binding sites are often relatively short (often 4-12, but up to 30 bases long), and not all positions in the sequence are interacting (strongly) with the transcription factor, some sequence variation in the motif is common. Therefore, the TFBM are usually provided as positional weight matrices, which describe the relative occurrences of each base for each position. This can be converted into a graphical representation, or sequence logo, where the size and order of the stacked letters (A,C,G,T) represents the relative occurrence of the base at that position (Figure 6). These motifs may be investigated for particular genes of interest to obtain a prediction on the Transcription Factor(s) that regulate their expression.

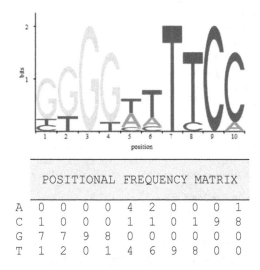

POSITIONAL FREQUENCY MATRIX									
A	0	0	0	0	4	2	0	0	0	1
C	1	0	0	0	1	1	0	1	9	8
G	7	7	9	8	0	0	0	0	0	0
T	1	2	0	1	4	6	9	8	0	0

Figure 6. The Transcription Factor Binding Motif for the NF-κB transcription factor *Relish / dorsal* of *Drosophila melanogaster*, depicted as sequence logo and Positional Frequency Matrix. The variation that is commonly found in the binding motif for a transcription factor is incorporated by specifying for each position in the motif the frequency at which each base is recorded. The size of the stacked letters for each position represent the relative occurrence of the respective bases on each position.

Apart from investigating the *cis*-regulatory elements for particular genes of interest, transcriptomics data is also highly suitable to test for over-represented *cis*-regulatory elements across (clusters of) co-expressed genes. This approach can identify groups of genes that are possibly co-regulated by the same Transcription Factor(s). Programs have been specifically developed to screen whether certain known motifs occur more often than you would expect by chance (for example, Clover [23]). These programs can also be extended with custom-made libraries of motifs, to include sequence motifs that could contain yet unidentified *cis*-regulatory elements. These novel motifs could be derived from aligning the upstream sequences of orthologs to identify conserved sequences among related taxa, or through the use of *de novo* motif discovery programs. Alternatively, MotifRegressor searches for any motif that is shared among genes that responded similarly in an expression study [24].

Analyzing the *cis*-regulatory elements in co-expressed genes can be used to start unravelling the genetic architecture of a trait. In our study for the response to parasite infection, we identified seven *cis*-regulatory elements that were over-represented among the differentially expressed genes, using a combination of MotifRegressor and Clover. Three of these motifs were TFBMs for transcription factors that were known to be involved in the immune responses (the GATA-factor *serpent*, the NF-ΚB *Relish/dorsal* and the Janus kinase *Stat92E*), while three others novel motifs were identified that have not yet been associated with any regulatory function. The expression levels of the transcription factor *serpent* was not changed after parasitation, which may appear counter-intuitive as the TFBM was over-represented in differentially expressed genes. Analysing the expression patterns of the clusters of co-expressed genes with the enriched TFBMs, however, and linking these to functional annotations for these groups of genes, suggested that this transcription factor was drawn away from it regular functions in development and metabolism (co-regulated genes with lower expression levels), towards the activation of the immune response (co-regulated genes higher expression levels) [5]. Additionally, we could hypothesize that the novel motifs may also be involved in coordinating the immune response to parasite infection. Using the cisRED database [25] as a first exploration of these novel motifs, two of these motifs were retrieved as a predicted regulatory element in the human genome sequences, including a hit in the upstream region of a known trans-activator of the MHC II complex (ZXDA). Although the functional characterization of the novel motif is still awaiting, these examples illustrate the complex genetic interactions that may coordinate the regulation of a trait.

Not only transcription is regulated through regulatory sequences associated with genes, translation into proteins is also partially coordinated by regulatory sequences. A rich world of small non-coding RNA molecules have been discovered since the start of the genomic era, which added a completely new dimension to the regulation of gene interaction networks [26]. One large class of these non-coding RNAs, the microRNAs, bind to the 3′ untranslated regions (3′ UTRs) of mRNAs, inhibiting their translation by polymerases and targeting the mRNAs for degradation. Several databases exist that link target genes or sequence motifs in the 3′ UTR to specific microRNAs. These tools are accessible through the microRNA database miRBase [27]. Associating microRNAs to the genes in a gene list could be achieved in an analogous manner as the association to the transcription factors: either by searching for known microRNA binding motifs within the 3'UTRs of differentially expressed genes, or by searching for any over-represented or conserved motifs in the 3'UTRs among the genes in the gene list and trying to associate those to microRNAs.

Another approach to analyse the genetic architecture for a trait is to make use of protein-protein interaction (PPI) network databases. These databases contain the known and predicted protein-protein association network, based on experimental approaches (e.g., two hybrid assays, purification of protein complexes, Chromatin immunoprecipitation (ChIP), etc.) and/or computational methods for predicting protein interactions. A large collection of these PPI databases is publicly available (see for example the Jena Protein-Protein Interaction website ppi.fli-leibniz.de/jcb_ppi_databases.html for an extensive overview). Several web-based tools can be used to analyse and visualize the PPI networks (e.g., STRING [28] and VisANT [29]). Gene lists submitted to these tools are being assembled into inter-connected networks of proteins, based on the PPI databases. The submitted proteins, as well as the proteins that it is known (or predicted) to interact with, form the 'nodes' in the network. All connections between any of these proteins (directly, or through an intermediary protein) are depicted by lines or 'edges' (Figure 7). The topology of these networks describe the frequency distributions of edges per node, and this can reveal whether the network resembles a random assembly of proteins or not [30].

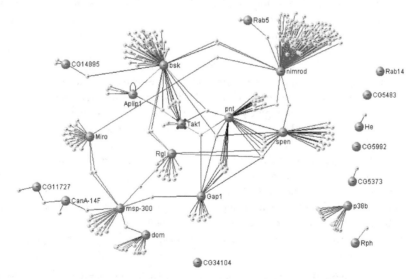

Figure 7. A Protein-Protein Interaction (PPI) network for a subset of the genes involved in the regulation of blood cell proliferation and differentiation in *Drosophila*. The proteins (or 'nodes') are depicted by red or blue circles. The red symbols represent genes with changed expressed in a *Drosophila* strain with an increased immunological resistance against parasites [6]. The known PPIs among these proteins are depicted by lines (or 'edges') between nodes, mostly based on two-hybrid data. Some of the proteins are highly interconnected to other modules of proteins (e.g., pnt, bsk), and these genes can be considered 'hubs' or key coordinators of the changes in expression.

Constructing a PPI network for genes that changed expression in a transcriptomics study may reveal modules of genes that are associated through functional processes, or identify key regulators/modulators to the treatment or condition of interest. Different than with the clustering of genes based on similarity of expression patterns for various conditions or time points, a PPI network will also group genes together that behaved very different transcriptionally, yet may participate in the same signal transduction cascade. We assembled a PPI network for the genes that changed expression between two *Drosophila* lines from the same genetic background, but differing genetically in their resistance to parasites after only five generations of artificial selection. Approximately a third of the nearly 900 genes with changed expression were inter-connected in several modules through an intricate and non-random PPI network [6]. Some genes could be identified within the network that had a central position with a high level of interconnectedness, and these genes may function as a 'hub', as they have the potential to influence the activity of a large number of genes. These 'hub' genes, or their regulators, could be hypothesized to provide targets for selection for increased resistance to parasites, in regulating and coordinating a multitude of phenotypic responses.

Another aspect of the genetic architecture of a trait is its relation to the genome architecture. The genes in a gene list can be mapped to chromosomal positions to search for chromosomal 'hotspots' of differential expression. Transcriptional activity varies for chromosomal domains or regions, and characterizing these patterns may indicate regulatory mechanisms that act on these genes. For example, some chromosomal domains are highly transcribed due to epigenetic mechanisms (e.g., chromatin architectures) that maintain a high activity state, as is seen for heat-shock genes [31]. Such domains under epigenetic control may be recognized by mapping multiple highly expressed genes, or conversely, a complete lack of expression, in the same chromosomal region. Such genomic domains may evolve at a different rate. For instance, the regions around heat-shock genes are more susceptible to insertion by Transposable Elements (i.e., mobile DNA sequences that can translocate themselves within the genome) due to their chromatin architecture, which may lead to a faster accumulation of mutations [32]. Furthermore, some chromosomal domains are highly transcribed in particular tissues only, and the gene arrangements within these domains are highly conserved across taxa [33]. Moreover, chromosomal regions show different expression patterns in healthy tissues compared to cancers [34]. These examples indicate that the physical arrangement of genes within the genome may be a target of evolution, likely due to epigenetic and other regulatory mechanisms that control gene expression of sections of the genome.

Additionally, examining the genomic positions of differentially expressed genes may reveal evolutionary processes that acted on the genes. Strong selection for a particular allele or genomic variant leaves a detectable pattern in the genome, which may be represented by a genomic clustering of genes with changed expression levels. When a particular allele provides an selective advantages to the individual, this locus may be swept through the population. Any allelic variation that is physically linked to this locus (i.e., resides in the nearby chromosomal domain) would be swept through the population as well. One of the best examples of a strong selective sweep is a mutation in the lactase gene in humans, that confers lactose resistance and is highly common among Europeans. Yet, not only this mutation has spread through the European population, but a region spanning approximately a million bases was swept along as well [35]. Such a sweep can also be detectable in expression assays. In our own studies, we imposed a strong selective sweep for immunological resistance in *Drosophila* against parasites, and mapped the genes with changed expression to the chromosomes. This revealed a part of one chromosome bearing a signature of positive selection [6].

Especially when information is available on sequence variations (i.e., different genotypes, or alleles) among the different biological samples in the experiment, genome-wide association mapping (GWAS) is another option to start unravelling the genetic architecture of a trait. In this approach, the individual variation in sequence is related to the variation in expression by statistical modelling. Using a multiple regression approach, the allelic states at various loci (e.g., whether it has an A, T, C or G at locus x, an insertion or deletion (indel), or inversion) is related to the expression level of each gene. This approach can be applied both when the sequence variation is independently acquired, for example through independent genotyping assays on the same samples, or from the more detailed information that can be extracted from tiling arrays or transcriptome sequencing data. This approach requires large sample sizes to obtain sufficient power and resolution for the statistical modelling, and has been used in a medical context to associate rare mutations with diseases. Causally linking sequence variants to diseases, however, has proved to be daunting [36]. Yet, this approach has been useful in obtaining more basal knowledge on genome functioning, and the relative importance of various sequence variants (e.g., copy number variants (CNVs), Single Nucleotide Polymorphisms (SNPs), small insertions and deletions (indels)) on gene expression variation [37].

5. Conclusions

Transcriptomics analysis has been hugely popular to explore the unknown players in a wide range of biological processes, diseases, traits and responses to stimuli. The technique is extremely powerful as a first step to implicate novel genes and pathways that may be involved or associated with a particular condition. It should be emphasized, however, that a difference in expression *per se* is not sufficient evidence to infer a direct involvement of the gene in the particular process or trait. This is a limitation of the technology, and it urgently requires the development of high-throughput empirical approaches to validate and functionally characterize the large numbers of genes that are putatively of interest. The availability of genome-wide libraries of RNAi stocks to knock down any gene of interest [38], or reference panels of genetic variants with fully sequenced genomes [39] are prime examples of the resources that are needed to follow up on transcriptomics studies. At the same time, the list of genes with potential involvement is certainly not the only information that can be derived from a transcriptomics study. It is especially the information on all the differentially expressed genes, including those that are not directly involved, that provides an exceptional source of information on regulation, correlated responses and the genetic architecture of a trait.

A large number of databases and bio-informatic tools are publically available to explore and annotate the individual genes on the gene list, and more importantly, to analyse the gene list collectively. The latter provides both additional power and a more comprehensive insight in the mechanisms and genetic architecture of a trait. Most traits, diseases and responses to environmental stimuli are highly complex, with environmental factors and genetic networks of interactions that contribute to the trait, disease or response. The factors and genetic network underlying a trait may be elucidated by a combination of bioinformatics approaches, and the emergent properties of such approaches may be more revealing than the search for individual candidates for a trait or process.

Many of the bio-informatic tools that can be applied for these analyses have been made accessible to the research community through the Bioconductor platform (www.bioconductor.org) [40]. This platform is based primarily on the open-source R programming language and runs on all operating systems. A good introduction into this versatile bio-informatic environment has been made available by the Girke lab at the University of California, Riverside through a combination of online manuals (http://manuals.bioinformatics.ucr.edu/). Other freely available, online suites for the analysis of transcriptomics data include Babelomics (http://www.babelomics.org) [41] and Galaxy (http://galaxy.psu.edu/, especially for transcriptome sequencing) [42-44].

The latest development in high-throughput sequencing are opening up new possibilities for the analysis of transcriptomics data. More detailed characterization of transcripts is achievable, and the power of transcriptomics analysis can now also be fully harnessed for organisms without a sequenced genome. Many of the approaches that have been developed for transcriptomics data with microarrays are equally applicable to data from transcriptome sequencing. In that sense, the knowledge-base that has accumulated in the research community in transcriptomics analysis over the past decade will largely remain a valuable resource. The experience and expertise that has been developed in dealing with the limitations and possibilities of analysing microarray data will also be of use while exploring the specific limitations and opportunities that are associated with this new platform. Robust and accurate methods need to be developed fast for the pre-processing, normalizing and analysing of transcriptome sequencing data. This will ensure that the full potential of this new technology can be made accessible to the wide research community.

Author details

Bregje Wertheim
Evolutionary Genetics, Centre for Ecological and Evolutionary Studies,
University of Groningen, Groningen, The Netherlands

Acknowledgement

I thank Eric Blanc and Eugene Schuster for their advice and our valuable discussions on the various bioinformatics approaches in gene expression studies. BW was supported by funding from the Netherlands Organization for Scientific Research (NWO) (Vidi grant no. 864.08.008).

6. References

[1] Hey Y, Pepper SD. (2009) Interesting times for microarray expression profiling. Brief Funct.Genomic Proteomic. 8(3):170-173.

[2] Mockler TC, Chan S, Sundaresan A, Chen H, Jacobsen SE, Ecker JR. (2005) Applications of DNA tiling arrays for whole-genome analysis. Genomics. 85(1):1-15.

[3] Wang Z, Gerstein M, Snyder M. (2009) RNA-Seq: a revolutionary tool for transcriptomics. Nat.Rev.Genet. 10(1):57-63.

[4] Ozsolak F, Milos PM. (2011) RNA sequencing: advances, challenges and opportunities. Nat.Rev.Genet. 12(2):87-98.

[5] Wertheim B, Kraaijeveld AR, Schuster E, Blanc E, Hopkins M, Pletcher SD, Strand MR, Partridge L, Godfray HCJ. (2005) Genome-Wide Gene Expression in Response to Parasitoid Attack in *Drosophila*. Genome Biology. 11(6):R94.

[6] Wertheim B, Kraaijeveld AR, Hopkins MG, Walther Boer M, Godfray HC. (2011) Functional genomics of the evolution of increased resistance to parasitism in Drosophila. Mol.Ecol. 20(5):932-949.

[7] Lemaitre B, Hoffmann J. (2007) The host defense of *Drosophila melanogaster*. Annual Review of Immunology. 25:697-743.

[8] Allison DB, Cui X, Page GP, Sabripour M. (2006) Microarray data analysis: from disarray to consolidation and consensus. Nat.Rev.Genet. 7(1):55-65.

[9] Hansen KD, Irizarry RA, Wu Z. (2012) Removing technical variability in RNA-seq data using conditional quantile normalization. Biostatistics. 13(2):204-216.

[10] Tsai CA, Wang SJ, Chen DT, Chen JJ. (2005) Sample size for gene expression microarray experiments. Bioinformatics. 21(8):1502-1508.

[11] Storey JD, Tibshirani R. (2003) Statistical significance for genomewide studies. Proceedings of the National Academy of Sciences USA. 100(16):9440-9445.

[12] Lathe W, Williams J, Mangan M, Karolchik D. (2008) Genomic Data Resources: Challenges and Promises. Nature Education. 1(3).

[13] Gotz S, Garcia-Gomez JM, Terol J, Williams TD, Nagaraj SH, Nueda MJ, Robles M, Talon M, Dopazo J, Conesa A. (2008) High-throughput functional annotation and data mining with the Blast2GO suite. Nucleic Acids Res. 36(10):3420-3435.

[14] Huang da W, Sherman BT, Lempicki RA. (2009) Bioinformatics enrichment tools: paths toward the comprehensive functional analysis of large gene lists. Nucleic Acids Res. 37(1):1-13.

[15] Breslin T, Eden P, Krogh M. (2004) Comparing functional annotation analyses with Catmap. BMC Bioinformatics. 5:193.

[16] Shah PK, Tripathi LP, Jensen LJ, Gahnim M, Mason C, Furlong EE, Rodrigues V, White KP, Bork P, Sowdhamini R. (2008) Enhanced function annotations for Drosophila serine proteases: A case study for systematic annotation of multi-member gene families. Gene. 407(1-2):199.

[17] Wang ET, Sandberg R, Luo S, Khrebtukova I, Zhang L, Mayr C, Kingsmore SF, Schroth GP, Burge CB. (2008) Alternative isoform regulation in human tissue transcriptomes. Nature. 456(7221):470-476.

[18] Haas BJ, Zody MC. (2010) Advancing RNA-Seq analysis. Nat.Biotechnol. 28(5):421-423.

[19] Krallinger M, Valencia A, Hirschman L. (2008) Linking genes to literature: text mining, information extraction, and retrieval applications for biology. Genome Biol. 9 Suppl 2:S8.

[20] Hoffmann R, Valencia A. (2004) A gene network for navigating the literature. Nat.Genet. 36(7):664-664.

[21] Chintapalli VR, Wang J, Dow JAT. (2007) Using FlyAtlas to identify better Drosophila melanogaster models of human disease. Nat Genet. 39(6):715.

[22] Maston GA, Evans SK, Green MR. (2006) Transcriptional regulatory elements in the human genome. Annu.Rev.Genomics Hum.Genet. 7:29-59.

[23] Frith MC, Fu Y, Yu L, Chen J-, Hansen U, Weng Z. (2004) Detection of functional DNA motifs via statistical over-representation. Nucleic Acids Research. 32.(4):1372-1381.

[24] Conlon EM, Liu XS, Lieb JD, Liu JS. (2003) Integrating regulatory motif discovery and genome-wide expression analysis. PNAS. 100(6):3339–3344.

[25] Robertson G, Bilenky M, Lin K, He A, Yuen W, Dagpinar M, Varhol R, Teague K, Griffith OL, Zhang X, Pan Y, Hassel M, Sleumer MC, Pan W, Pleasance ED, Chuang M, Hao H, Li YY, Robertson N, Fjell C, Li B, Montgomery SB, Astakhova T, Zhou J, Sander J, Siddiqui AS, Jones SJ. (2006) cisRED: a database system for genome-scale computational discovery of regulatory elements. Nucleic Acids Res. 34(Database issue):D68-73.

[26] Zamore P, Haley B. (2005) Ribo-gnome: The big world of small RNAs. Science. 309(5740):1519-1524.

[27] Griffiths-Jones S, Saini HK, van Dongen S, Enright AJ. (2008) miRBase: tools for microRNA genomics. Nucleic Acids Res. 36(Database issue):D154-8.

[28] Szklarczyk D, Franceschini A, Kuhn M, Simonovic M, Roth A, Minguez P, Doerks T, Stark M, Muller J, Bork P, Jensen LJ, von Mering C. (2011) The STRING database in 2011: functional interaction networks of proteins, globally integrated and scored. Nucleic Acids Res. 39(Database issue):D561-8.

[29] Hu Z, Hung JH, Wang Y, Chang YC, Huang CL, Huyck M, DeLisi C. (2009) VisANT 3.5: multi-scale network visualization, analysis and inference based on the gene ontology. Nucleic Acids Res. 37(Web Server issue):W115-21.

[30] Jeong H, Tombor B, Albert R, Oltvai ZN, Barabasi AL. (2000) The large-scale organization of metabolic networks. Nature. 407(6804):651-654.

[31] Farkas G, Leibovitch BA, Elgin SC. (2000) Chromatin organization and transcriptional control of gene expression in Drosophila. Gene. 253(2):117-136.

[32] Walser JC, Chen B, Feder ME. (2006) Heat-shock promoters: targets for evolution by P transposable elements in Drosophila. PLoS Genet. 2(10):e165.

[33] Yamashita T, Honda M, Takatori H, Nishino R, Hoshino N, Kaneko S. (2004) Genome-wide transcriptome mapping analysis identifies organ-specific gene expression patterns along human chromosomes. Genomics. 84(5):867-875.

[34] Caron H, van Schaik B, van der Mee M, Baas F, Riggins G, van Sluis P, Hermus MC, van Asperen R, Boon K, Voute PA, Heisterkamp S, van Kampen A, Versteeg R. (2001) The human transcriptome map: clustering of highly expressed genes in chromosomal domains. Science. 291(5507):1289-1292.

[35] Bersaglieri T, Sabeti PC, Patterson N, Vanderploeg T, Schaffner SF, Drake JA, Rhodes M, Reich DE, Hirschhorn JN. (2004) Genetic signatures of strong recent positive selection at the lactase gene. Am.J.Hum.Genet. 74(6):1111-1120.

[36] Marian AJ. (2012) Molecular genetic studies of complex phenotypes. Translational Research. 159(2):64-79.

[37] Stranger BE, Forrest MS, Dunning M, Ingle CE, Beazley C, Thorne N, Redon R, Bird CP, de Grassi A, Lee C, Tyler-Smith C, Carter N, Scherer SW, Tavaré S, Deloukas P, Hurles ME, Dermitzakis ET. (2007) Relative impact of nucleotide and copy number variation on gene expression phenotypes. Science. 315(5813):848-53.

[38] Dietzl G, Chen D, Schnorrer F, Su KC, Barinova Y, Fellner M, Gasser B, Kinsey K, Oppel S, Scheiblauer S, Couto A, Marra V, Keleman K, Dickson BJ. (2007) A genome-wide transgenic RNAi library for conditional gene inactivation in Drosophila. Nature. 448(7150):151-156.

[39] Mackay TF, Richards S, Stone EA, Barbadilla A, Ayroles JF, Zhu D, Casillas S, Han Y, Magwire MM, Cridland JM, Richardson MF, Anholt RR, Barron M, Bess C, Blankenburg KP, Carbone MA, Castellano D, Chaboub L, Duncan L, Harris Z, Javaid M, Jayaseelan JC, Jhangiani SN, Jordan KW, Lara F, Lawrence F, Lee SL, Librado P, Linheiro RS, Lyman RF, Mackey AJ, Munidasa M, Muzny DM, Nazareth L, Newsham I, Perales L, Pu LL, Qu C, Ramia M, Reid JG, Rollmann SM, Rozas J, Saada N, Turlapati L, Worley KC, Wu YQ, Yamamoto A, Zhu Y, Bergman CM, Thornton KR, Mittelman D, Gibbs RA. (2012) The Drosophila melanogaster Genetic Reference Panel. Nature. 482(7384):173-178.

[40] Gentleman RC, Carey VJ, Bates DM, Bolstad B, Dettling M, Dudoit S, Ellis B, Gautier L, Ge Y, Gentry J, Hornik K, Hothorn T, Huber W, Iacus S, Irizarry R, Leisch F, Li C, Maechler M, Rossini AJ, Sawitzki G, Smith C, Smyth G, Tierney L, Yang JYH, Zhang J. (2004) Bioconductor: open software development for computational biology and bioinformatics. Genome Biology. 5:R80.

[41] Medina I, Carbonell J, Pulido L, Madeira SC, Goetz S, Conesa A, Tarraga J, Pascual-Montano A, Nogales-Cadenas R, Santoyo J, Garcia F, Marba M, Montaner D, Dopazo J. (2010) Babelomics: an integrative platform for the analysis of transcriptomics, proteomics and genomic data with advanced functional profiling. Nucleic Acids Res. 38:W210-W213.

[42] Blankenberg D, Von Kuster G, Coraor N, Ananda G, Lazarus R, Mangan M, Nekrutenko A, Taylor J. (2010) Galaxy: a web-based genome analysis tool for

experimentalists. Current protocols in molecular biology / edited by Frederick M.Ausubel ...[et al.]. Chapter 19:21.

[43] Goecks J, Nekrutenko A, Taylor J, Galaxy Team. (2010) Galaxy: a comprehensive approach for supporting accessible, reproducible, and transparent computational research in the life sciences. Genome Biol. 11(8):R86.

[44] Giardine B, Riemer C, Hardison RC, Burhans R, Elnitski L, Shah P, Zhang Y, Blankenberg D, Albert I, Taylor J, Miller W, Kent WJ, Nekrutenko A. (2005) Galaxy: A platform for interactive large-scale genome analysis. Genome Res. 15(10):1451-1455.

Chapter 4

The REACT Suite: A Software Toolkit for Microbial *RE*gulon *A*nnotation and *C*omparative *T*ranscriptomics

Peter Ricke and Thorsten Mascher

Additional information is available at the end of the chapter

1. Introduction

The 'age of omics' has provided a wealth of genomic and transcriptomic information that is readily available in public databases. In September 2011, GOLD (the Genomes OnLine Database)[1] (Liolios *et al.*, 2010) lists more than 2,600 finished microbial genome sequences and more than twice this number for ongoing and incomplete genome projects, not counting the plethora of metagenome projects, which provide even larger sequence compilations. Comparable numbers of datasets can be retrieved from the two major microarray databases, the Stanford Microarray Database (SMD)[2] (Demeter *et al.*, 2007), and the Gene Expression Omnibus (GEO database)[3] (Barrett *et al.*, 2011) hosted by the National Center for Biotechnology Information (NCBI), which in September 2011 together provide over 9,000 bacterial microarray datasets to the public.

This enormous amount of data provides a treasure chest of information ready to explore. In recent years, a number of powerful comparative genomics databases such as GenoList[4] (Lechat *et al.*, 2008), or MicrobesOnline[5] (Dehal *et al.*, 2010) have provided the community with toolkits to make use of this information.

Combining microarray data with genomic information is a particular powerful approach for identifying and predicting regulons, which are regulatory units consisting of a number of genes or operons under the control of specific transcription factors. Such studies require the

---

[1] URL for the GOLD database: http://genomesonline.org
[2] URL for the Stanford Microarray database: http://smd.stanford.edu
[3] URL for the Gene Expression Omnibus: http://www.ncbi.nlm.nih.gov/geo/
[4] URL for GenoList: http://genolist.pasteur.fr/
[5] URL for the MicrobesOnline database: http://www.microbesonline.org

identification of co-expressed genes (indicative of co-regulation) from in-depth comparative transcriptome profiling, combined with genomic information, including operon structure, genomic context conservation and the presence of specific regulator binding sites.

The major problem of combining genomic with transcriptomic data to ultimately extract meaningful regulon information is the lack of defined standard formats and software interfaces that allow a direct transfer of data sets derived from transcriptome analyses to comparative genomics databases and vice versa. The REACT suite was developed with the purpose in mind to facilitate such combinations of the different analysis steps outlined above in one intuitive and user-friendly environment. Transcriptome datasets from different sources can be integrated into REACT via a sophisticated import interface and are stored, together with the cognate genomic information, in a MySQL database. This database, together with the central part of the software toolkit and all interlinked third-party tools run on a central computer, which actually performs the analyses: the "REACT-server". It is accessed by the user-interface ("REACT-client") via inter- or intranet. The user will solely work with the corresponding client program, which can be installed on the personal computers or laptops of various users. While the installation of the REACT-server demands some technical knowledge, the client can be run easily on computers with a java runtime environment.

Taken together, the REACT suite provides users with a simple-to-use but powerful bioinformatics environment to perform regulon annotation and comparative genomics analyses based on microarray data and genome sequences. Both server and client software of the REACT suite are freely available from the corresponding author.

2. The basic concept of REACT

REACT was developed to enable users to perform the various steps of expression- and regulon analyses in a quick and intuitive manner. Tools are no longer separated entities demanding different and often incompatible data formats, but can be rather regarded as parts of a comprehensive, fully integrated unit. Data from a wide range of sources can be collected and analysed together. When working with REACT, the user has access to the various representations of the data as well as to the analysis tools via so-called "views" that are intuitively interlinked to enable an interactive flow of both data and analyses:

The "GeneView" displays gene-centric information including DNA- and amino acid sequences, links to a number of external databases, as well as the genomic context of a gene in the form of a simple genome browser.

The "RegulonView" lists all genes controlled by the same regulator as well as binding motif(s), individual binding sites, alternative promoters etc., based both on the information stored in curated public regulon databases and data added by REACT users.

The "ArrayView" allows both importing new microarray datasets and performing data analysis on existing datasets. REACT has a sophisticated interface for the import of array data in nearly every tabulated data format from individual proprietary formats up to GEO/SMD datasets. Data analysis includes one- and two-dimensional scatterplot analyses of signal or ratio-values, as well as gene- and array-clustering with various hierarchical clustering methods, distance methods and correction algorithms (normalization, gene-centering, log-transformation) of all or only selected (collected) subsets of the data.

The *"MotifView"* contains the information of all sequence motifs of known or putative regulator binding sites collected in the current REACT database. Moreover, it enables users to perform MEME analyses to discover new regulatory elements in the upstream regions of selected annotated genes or operons and MAST analyses of previously computed or imported motifs against pre-compiled upstream sequence datasets.

The concept of REACT includes an in-depth integration of the different views via links, enabling users to switch easily between different aspects of the data. Most views are flexible and can be extended with additional data fields to accommodate additional external links, allowing more individualized views and analyses of the data.

Moreover, wherever gene or array data are displayed, the user can easily collect them, thereby creating a data subset available as input for all other implemented analyses. During the various analysis steps, these collections can be continuously changed and expanded, again by selecting single genes and arrays or whole groups of them, such as groups of genes clustering together within a scatter plot analysis. All collected or "marked" arrays and genes are displayed throughout the various views of REACT in form of sortable lists. The items of these lists act as internal links to the corresponding detailed *Array-* or *GeneViews*. Current collections can be saved and opened again for later use, so that the user can easily switch between different data sets any time. In addition, the sequences of the selected genes can also be exported into external FASTA files.

The implemented REACT-databases are organism-specific. In its current version, REACT contains two databases for the model bacteria *Escherichia coli* and *Bacillus subtilis*, but could also be extended to other microbial species. Each REACT-database is based on the detailed genomic data of the model organism, which will be described in the following paragraphs, as well as of an extendable amount of microarray data of this organism. Moreover, basic genomic information on related organisms, the so-called "reference organisms" is also integrated and can be included into some of the analyses. The list of reference genomes can easily be extended, to adjust a given database to the dynamics in genome sequencing.

3. Description of the individual views of the REACT suite

The information stored in REACT databases can be accessed via so-called views that display the data, allow their selection and provide functional links between different types of data for their interactive analysis. In the following sections, we will describe the major views of REACT, to provide an overview of their features.

3.1. The *GeneView*

The *"GeneView"* bundles all available information about individual genes (Fig. 1). On top of the page, the gene identifiers are displayed, accordingly to the existing genomic nomenclature. The first identifier is the gene name, e.g. *"icd"* in case of the *B. subtilis* isocitrate dehydrogenase. If more than one gene name exists for a given gene, the nomenclature applied by REACT is derived from the genome annotation stored in the NCBI genome database[6]. Here, as in other views of the REACT suite, active features working as internal links are highlighted by red letters (Fig. 1). In case of gene names, a double click would bring the user to the corresponding *GeneView*, while a single click would mark the gene (= add it to the gene collection in the left panel) for further analyses.

Figure 1. The *GeneView*. Exemplary screenshot for the gene *icd*, encoding the isocitrate dehydrogenase. See text for details.

In addition to the name, each gene has a unique gene-ID or gene number, which consists of an abbreviation for the organism and a number of the gene (based on the chromosomal position). For example, the identifier of the *B. subtilis* isocitrate dehydrogenase is "BSU29130" (Fig. 1). Gene names and numbers are the major identifiers that are used throughout the several displays of the REACT suite. The gene numbers cannot be modified by the users to ensure the integrity of the database. The putative or known functions of the encoded proteins are shown below the gene name, including synonyms and alternative descriptions (if present).

The central part of the *GeneView* are the external links and descriptive data fields. For each genome database implemented in the REACT suite, links to important public databases are already predefined for each gene. This include links to the COGs (Cluster of Orthologous Genes) database[7] (Tatusov *et al.*, 1997), hosted again by the NCBI, the Enzyme Nomenclature

[6] URL for the NCBI genome database: http://www.ncbi.nlm.nih.gov/sites/genome
[7] URL for the COGs database: http://www.ncbi.nlm.nih.gov/COG

database[8] (Bairoch, 2000), the already mentioned MicrobesOnline comparative genomics database, the NCBI Protein database[9], the protein data bank PDB[10], a collection of protein structures and structure-related information (Rose et al., 2011), the Pfam[11] (Finn et al., 2010), Prosite[12] (Sigrist et al., 2010) and SMART[13] (Letunic et al., 2009) databases, all of which are dedicated to the definition, maintenance and easy identification of protein domains and families. Further predefined links include a link to Pubmed and to Google. In addition to these general sites, the *GeneView* page also links to organism-specific databases and genome resources, such as BSORF[14] or SubtiList[15] in case of *B. subtilis*.

For all of the above, the links in the REACT databases are gene-specific and directly connect the user with the cognate gene/protein-specific page of the external database. Depending on the type of the external database and the information available for the displayed gene, zero to many external hits will be provided as links. If no such specific database identifier exists, as in the case of Google's search engine, a gene-related term (e.g. the gene name) has been chosen as the link parameter. REACT is highly adjustable to the individual users' needs. Hence, the external links are not limited to those preimplemented in the existing REACT databases for *E. coli* and *B. subtilis*. (see section 5.3 "Modifying REACT: the administrator mode" for details).

In addition to the links and data fields, the *GeneView* also displays the DNA and amino acid sequences of the current gene, which are linked to the BLASTn and BLASTp tools[16] at NCBI. The user is therefore able to directly search for similar sequences in the public domain. Moreover, a genome browser is implemented at the bottom of the *GeneView* for a quick glance on the genomic environment of the current gene (Fig. 1). The gene icons are coloured according to the COG-functional classes assigned to each gene and serve as links to the corresponding *GeneViews*.

Two additional functions are available in the *GeneView*. First, the user can retrieve the upstream genomic region of the gene via a specific dialog box, based on user-provided information, such as upstream region length, inclusion of start codon, or choice between the upstream region of the current gene or the first gene of its operon. The latter function is very useful for collecting upstream regions for motif searches (see section 3.5). Second, expression data of the active gene can directly be retrieved from the REACT microarray database. For this, the user can choose either all or only selected microarray datasets, and limit the set of extracted values by a certain threshold expression ratio level. The *GeneView* is therefore not only the central platform for all gene-centric data, but is also directly linked to all other views described in the following sections.

---

[8] URL for the ENZYME database: http://enzyme.expasy.org/enzyme_ref.html
[9] URL for the NCBI Protein database: http://www.ncbi.nlm.nih.gov/protein
[10] URL for PDB: http://www.rcsb.org/
[11] URL for the Pfam database: http://pfam.sanger.ac.uk/
[12] URL for the Prosite database: http://prosite.expasy.org/
[13] URL for the SMART database: http://smart.embl-heidelberg.de/
[14] URL for the BSORF site: http://bacillus.genome.ad.jp/
[15] URL for SubtiList: http://genolist.pasteur.fr/SubtiList/
[16] URL for BLASTn and BLASTp: http://blast.ncbi.nlm.nih.gov/Blast.cgi

3.2. The *OperonView*

Operons are transcriptional units consisting of two or more neighbouring genes that are co-expressed. If a gene has been assigned to an operon and annotated accordingly in the REACT database, a link from the *GeneView* leads to the *OperonView*. Both views are organized in a similar fashion and the *OperonView* also contains a genome browser. It is identical to the *GeneView*'s with the exception that here the current operon is highlighted.

The operon identifier is again immutable, since it is used by REACT as internal reference. The operon name by default consists of the concatenated names of the genes within this operon. When displayed outside of the *OperonView*, it functions as an internal link, enabling the user to jump directly to the corresponding *OperonView*. From within the *OperonView*, it can be used as a link to an external database containing additional information regarding this operon.

In addition to providing a direct link to all corresponding *GeneView* pages, the *OperonView* also provides a list of and links to all regulons, to which the current operon belongs. They are represented by their REACT-internal regulon identifier, the name of the corresponding transcription factor and a brief description of the regulon (see the following section for details).

3.3. The *RegulonView*

The next higher level of genetic units is the regulon, which consists of a number of genes or operons under the direct control of a specific transcription factor. Regulons are displayed within the REACT suite in the *RegulonView*, which is subdivided into two panels. The first section ("All regulons") contains a tabulated list of all regulons currently defined in the REACT database, which includes the most important information such as the regulon-ID, the main description of the regulon and the associated transcription factor. The second part displays the detailed view of a specific regulon selected from the first list ("Act. regulon"). This view is organized similar to the *GeneView* or *OperonView*. It contains regulon identifier, a link to the corresponding transcription factor (if known) and the sequence motif of its cognate DNA-binding site, which is found upstream of the regulated operons or genes comprising this regulon.

The regulon-ID is derived from the gene-ID of the corresponding transcription factor and marked by the extension "_R". It is implemented as an active link that directly connects to an external regulon database. In case of *B. subtilis*, this is primarily DBTBS[17], the database of transcriptional regulation in *B. subtilis* (Sierro *et al.*, 2008), but BSORF or SubtiList have also been used for the initial regulon annotation. For *E. coli*, the regulon information has been extracted from RegulonDB[18] (Gama-Castro *et al.*, 2011). Additional regulon definitions can be added at any time, including putative regulons with only rudimentary information, such

---

[17] URL for the DBTBS database: http://dbtbs.hgc.jp/
[18] URL for RegulonDB: http://regulondb.ccg.unam.mx/

as sets of co-expressed genes. If the regulator DNA-binding site is known and defined, it will also be displayed as a so-called SequenceLogo (see section 3.5 for details). Below the general regulon-associated information and, if available, the sequence motif overview, additional data fields and external links are displayed.

The central part of the *RegulonView* provides individual information on all associated operons and genes, including the name, the first gene in case of operons, the corresponding σ factor and the position and sequence of the putative binding site in front of the regulated transcriptional units. This information enables the user to get a quick first impression of the regulated genes of this regulon.

3.4. The *ArrayView*

All three views explained so far are highly similar to one another and strongly integrated, not only regarding the information provided but also in the way the user can navigate from one view to the next. They all provide a gene-centric view on the REACT data and invariantly rely on the genomic sequence as a reference. Regulons consist of operons, which are made up of individual genes with a defined position on the chromosome. The same is true for regulator binding sites.

In contrast, the *ArrayView* provides access to the second central data pool stored within the REACT database, the microarray datasets. Array data exist in a great variety of different formats. Especially data sets from the early years of transcriptome studies often are available only in form of simple tables or excel spread sheets without any defined data format, distributed over numerous journal homepages or webpages of individual research groups, making their implementation into comparative transcriptome analysis very difficult. As a result, public databases, such as the already mentioned GEO or SMD, have been developed for the storage and description of microarray datasets that comply to the MIAME (Minimal Information About a Microarray Experiment)[19] standard (Brazma *et al.*, 2001). Unfortunately, these databases still contain only a fraction of the published microarray datasets. The biggest challenge for a comparative transcriptome database is therefore to organize and import microarray data from diverse sources into a compatible format.

3.4.1. Organizing microarray data in the REACT suite

A complete microarray dataset contains at least three types of information. (i) A list of all genes represented by a given DNA microarray, which is linked to the corresponding expression values, either expressed as (ii) raw fluorescence values for the reference and experimental condition, or as (iii) the respective ratio (or fold-change) between the two conditions. Within the REACT suite, such a data collection is called an "Array". Obviously, the Array is only useful if additional descriptive information (meta-information) is available. This can be a short description of the specific experimental set-up or a link where this information is stored. Often, a group of array datasets are related to each other and

---

[19] URL for the description of the MIAME standard: http://www.mged.org/Workgroups/MIAME/miame.html

described in a single format, e.g. as a result of one experiment. This is reflected by the REACT data format "Array Set".

The *ArrayView* is split into four sub-views. The first sub-view, "All Arrays" contains a tabulated view of all array sets of the current REACT database. If one array set is selected, all arrays of this set are displayed below the upper window, again in tabulated format. Both tables contain some basic meta-information on either the array or array set, respectively.

Selection of one array or array set leads to the next sub-view "Act. Array", which provides the detailed information, including the ID, name, a description of the underlying experiment, the source of the data, available literature, and external links. The "Array Set" subview lists all individual arrays within the set, which can be marked separately for further analysis. The most important feature of this sub-view is a tabulated, sortable list of all genes, for which data are available within this array. It contains information on the gene name, the signal value, the control value, the ratio of signal to control, the number of replicates that were combined, the arithmetic mean and error of the values. This data is normally directly derived from the original data sets. Two additional columns indicate which genes are currently marked and if their value can be trusted. The trust value is a simple way to allow users to flag single values as untrusted, thereby automatically excluding them from subsequent analyses. Trust values can be easily set for marked genes within the current subview.

The data table is sortable based on any column, e.g. high or low signals or ratio values. Genes of interest can be collected as "marked genes" for inclusion into follow-up analyses. Each gene-specific data row of the table functions as an internal link to the corresponding *GeneView*, thereby providing a direct connection between the array-centric data of the *ArrayView* and the gene-centric data of the *Gene/Operon/RegulonViews*.

An additional feature of the *ArrayView* is the "Similar gene" function. For each array displayed in the dialog, the user can define ratio-thresholds similar to the ratio of the current gene. REACT then automatically retrieves a list of all genes fulfilling the user defined criteria. These genes can then be marked for subsequent analyses. The "Similar gene" function therefore provides a simple but efficient and direct way to find genes with similar expression characteristics from the available microarray database.

3.4.2. Importing microarray data into the REACT database

As already mentioned, one of the major problems in comparative transcriptome analyses is the lack of a mandatory gold standard for array datasets, especially from the early, pre-MIAME era. But even ten years after this standard has been introduced, this problem is still far from be solved, and the number of microarray datasets not complying to these standards is still rising (Brazma, 2009).

Even implementing the minimum amount of information needed to integrate an array data set into the REACT database – a two-column table, with one column containing the gene identifiers and the second containing either the signal values or expression ratios between signal and control – can be daunting. Gene identifiers are either not used consistently (as synonyms often exist), or the DNA microarray might not contain all genes, or duplication of some. Likewise, signals can be represented as raw fluorescence values, either as mean or average values, in which case control values need to be provided or defined. Alternatively, a table might provide ratio of signal to control, which can be either expressed as log-values or as fold-changes. To facilitate handling and import such diverse types of data, the REACT suite contains an easy-to-use microarray import interface (Fig. 3).

During import, microarray data in any tabulated format is initially pre-loaded into the REACT import panel. REACT automatically detects the number of columns in the file and generates an adequate number of numerated preview columns for easy identification. After semiautomated discrimination of commentary lines, the appropriate type of information has to be assigned to each column. REACT needs at least one column containing the gene identifiers and one column for the signal or ratio values. Other types of information can also be assigned, such as the signal background, the control value, and the control background. Based on the assignment, REACT 'knows' what to do with the individual data, e.g. if background columns are specified, their values will be subtracted from the corresponding signal or control values. Ratios between signal and control values can either be directly imported or will be calculated, depending on the data provided. It is even possible to import data with only a single column containing the signal values (e.g. during time course experiments). In a later step, one of the imported arrays (e.g. time 0) can then be used as a standard control for all datasets to calculate the ratio values needed for most analyses. Large datasets containing many replicates of one experiment can be imported in a single table. In this case, REACT offers the possibility to average the sets of columns assigned for signal, control or ratio values.

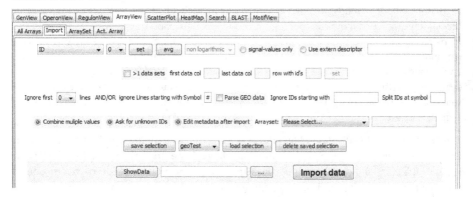

Figure 2. The microarray import interface of the *ArrayView*. See text for details.

If large numbers of different experiments are stored in a single table, they can be parsed at once using the "batch"-import. The user defines the different ratio-columns, and each column will be treated as a separate array, within a common array set. Moreover, it is possible to define, if ratio data are in logarithmic format (they will be converted to internal non-logarithmic values) or not.

One major challenge when comparing data from different sources and hence formats is dealing with variations and differences in the gene identifiers used in different microarray templates. REACT knows a large amount of different gene descriptions, as mentioned in section 3.1. During data import, REACT will accept any of these names and synonyms. But if unknown identifiers occur or synonyms have been assigned more than once in a microarray dataset (e.g. in case of different probes representing a single gene), REACT will ask the user for a specific decision. The user can then skip/delete the line, manually assign a gene name, or add the new synonym to the database for future use.

Taken together, REACT should be able to import virtually all formats of array data, as long as they are tabulated. For the more complex datasets, such as those generated by the GEO, special parsing options for the corresponding meta-information are available in REACT.

3.5. The *MotifView*

The *MotifView* is divided into five sub-sections, three of which (the "Upstream"-panel, the "Act. Motif" and the "MotifTable") are used for displaying the data and will be described here. The remaining two – MEME- and MAST-panel – are interfaces for the eponymous external analysis tools and will be discussed in more detail later (section 4.4).

The "Upstream"-panel is used to collect and display DNA regions upstream of coding sequences. Mostly, this will be intergenic regions, which are of particular interest, since they contain both (alternative) promoters and putative DNA-binding sequences of transcriptional regulators. The possibility to retrieve and manage such upstream regions is therefore of crucial importance in the context of regulon analyses. Upstream regions can be added to the "Upstream"-table by one of three means: (i) collectively from the active list of marked genes, (ii) individually by gene name, or (iii) directly from within the *GeneView* for the corresponding gene. In all cases, the user can define parameters for the retrieval, such as the sequence length, inclusion of sequence up to the upstream stop codon or exclusion of sequences of upstream genes, in the case that they overlap with the selected upstream sequence length.

The "Upstream"-table displays all upstream regions collected by the user in the course of an analysis by any of the three methods described above. For each upstream region, the ID and name of the corresponding gene, and the sequence and position of the respective region in the genome are displayed. These regions (or subsets thereof) can easily be removed or added, exported as FASTA-formatted sequence files or selected for further analyses, such as the MEME/MAST analyses (see section 4.4).

In the context of regulon analyses performed within the REACT suite, motifs are defined as short stretches of nucleotide sequence that are conserved in a collection of upstream regions, derived e.g. from co-expressed genes. They are expressed as so-called position-specific scoring matrices (PSSMs, also known as Position Weight Matrices, PWM) or regular expressions (RE), which both describe the probability for specific bases to occur at a specific position of the motif. Such matrices are graphically displayed as so-called "SequenceLogos", in which the height of the letters representing the four bases is a measure for the degree of conservation at any given position within the motif.

In REACT, defined motifs of known regulator-binding sites are stored in the "MotifTable". In this table, each motif is represented by the REACT-internal ID, the name of the motif (normally equivalent with the name of its cognate regulator), the motif length, the associated regular expression or PSSM, as well as the corresponding SequenceLogo. Selection of a motif opens the "Act. Motif"-panel, which provides all available information of one motif, including the name of the regulon it is associated with. This regulon name serves as an internal link to the corresponding page within the *RegulonView*. Moreover, a multiple sequence alignment of all (upstream) sequences underlying this motif is shown (if available), which can be exported as FASTA format.

4. Search options and analysis tools within the REACT suite

So far, this book chapter has described the major views that represent and display the data stored within the organism-specific REACT databases. In the following sections, we will describe the tools that allow the user to search the database and analyse genes, motifs, and microarray datasets in order to extract and define regulons. These tools include a search engine, an internal BLAST tool, cluster analysis and scatter blot tools for microarray datasets, as well as the MEME/MAST algorithms to identify and search for regulator binding sites in upstream regions of co-expressed genes.

4.1. The *Search* tool

The wealth of information stored in the REACT databases requires search tools to find specific data sets. The REACT Search tool contains four panels, enabling the user to search for genes, regulons, arrays and array-sets, respectively. These panels share the same general structure and differ only in minor features. The common features will be described for the gene search panel (Fig. 3).

Genes of interest can be searched by all gene-specific data fields, e.g. by gene-ID, name, synonyms, function, comments, but also any other user-defined field. These fields can be searched by a number of search strings, such as <containing>, <being equal to> or <starting with> a certain term. After the search hits are displayed in tabulated form in a result window below the search panel, where they can be marked or used as internal links to the respective views. Consecutive searches can be combined by <add>, <remove>, <keep> results or <negate> operations, thus enabling even for more sophisticated searches.

The search functions introduced so far are available in all four search panels. For genes and arrays, an additional function allows searching marked genes or arrays, respectively. Moreover, genes can also be successively searched by COG categories and COG terms.

Figure 3. The *Search* tool of REACT, exemplified by the search panel for genes.

4.2. The internal *BLAST* tool

Within the REACT suite, BLAST analyses (Altschul *et al.*, 1990) can be performed in two different ways. First, it can be performed from within the *GeneView* via a direct external link to the NCBI BLAST server (see section 3.1). Second, REACT also provides an internal BLAST search, which allows comparing a gene of interest with the internal reference genomes of the corresponding REACT database. This internal search, which can be accessed by the corresponding BLAST panel, allows retrieving not only the homologous gene or protein sequences, but also the corresponding upstream regions for further analyses, such as MEME/MAST (see section 4.4).

Both external (pasted into the input window) and internal (derived from the gene/protein displayed in the current *GeneView*) sequences can be used as query, either as DNA or protein sequence. After choosing the appropriate BLAST algorithm and the sequence data to be analysed, the results are displayed in tabulated form in the corresponding panel. For each match, both gene-specific information (ID, name, function, organism) and BLAST-specific values (E-value, per cent identity, match length, number of mismatches/insertions/ deletions) are displayed. Moreover, the genomic context is illustrated in a genome browser.

For each match, the DNA or amino acid sequence can be retrieved. Moreover, REACT also provides access to the corresponding upstream region via the "Retrieve upstream" function. The corresponding sequences will then be added to the "Upstream"-panel of the *MotifView* as described above (see section 3.5).

4.3. Microarray analysis tools

As mentioned before, the REACT suite is based on organism-specific databases that contain two types of data. The gene-centric data is derived from public genome sequence information and accessible through the *Gene-, Operon-, Regulon-*, and *MotifView*, while the array-centric data is displayed in the *ArrayView*. Two different types of tools have been implemented into the REACT suite in order to analyse this second type of data: (i) Scatter plot analyses (4.3.1) allow the comparison of up to two experimental conditions, while Cluster analyses (4.3.2) are used to extract expression values from multi-array comparisons.

4.3.1. The scatterplot tool

A scatter plot is a graphical way to project values for two variables of a data set into a two-dimensional grid, thereby placing similar samples in the same regions of the grid. The data is displayed as a collection of points, each having the value of one variable determining the position on the x axis and the value of the other variable determining the position on the y axis (Utts, 2005). A scatter plot is a very useful tool to identify similarities and differences in large, comparable datasets that agree in large parts with each other. The more the two data sets agree, the more the scatter tends to concentrate in the vicinity of a so-called identity line, where $y = x$.

Within REACT, scatter plot analyses are normally used to display genes according to their expression data of two selected arrays, using the expression value of the first array as x coordinate and the values of the other as y coordinate. This representation of the data results in an interactive panel where genes with similar expression patterns are grouped together.

In most cases, the vast majority of analysed genes should show the same expression values/ratios under both conditions and will therefore be placed closely together on the $x=y$ line. In contrast, genes that differ significantly in their behaviour between the two conditions will appear as outliers and can therefore be easily identified in the plot. Of course, comparisons of array datasets from different research groups tend to deviate more or less significantly from this ideal situation. Hence, the differences in experimental conditions need to be kept in mind when comparing array data sets.

Scatter plot analyses can be performed using either signal or ratio array values, thereby allowing to compare the behaviour of genes in the presence of different stimuli (ratio data), but also to compare different time points from one time course experiment (using signal data). Such comparisons of expression data from two different microarrays are called two-dimensional scatter plots (see Fig. 4 for an example).

But the user can also compare the data of one array against itself, using the same signal or ratio values as coordinates for the x and y axes. As this results in the placement of all genes on one line (the identity line), it is called a one-dimensional scatter plot. Such an analysis can be helpful to verify that a group of related genes (e.g. from one operon) behaves in a similar fashion within one experiment.

The input (expression data) for both types of analyses can either be log-transformed or normalized for the arrays or for the genes (array- and gene-centering, respectively). Moreover, the data can be filtered to remove "untrusted" genes prior to the analysis. Here, REACT removes all genes previously flagged as untrusted and un-reliable (either automatically during the import or later by the user) in one or both array datasets.

The major advantage over using external standard scatter plot tools is the deep integration of the REACT scatter plots with the REACT database. Without pre-selection of genes, the analysis will be carried out with the complete microarray data sets. Genes that specifically respond to only one of the two conditions will appear as outliers and can then be easily selected directly from the plot and thereby added to the list of marked genes directly for further analyses within REACT. This deep integration and direct connection of array-centric results with gene-centric information is one of the major strengths of the REACT suite, which enables the user to efficiently analyse even complex datasets.

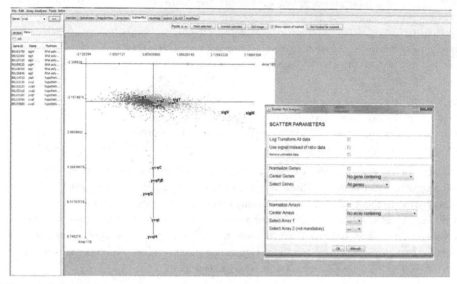

Figure 4. Example of a two-dimensional scatter plot analysis with labelled genes. The inset on the right shows the parameter window for choosing the settings for a scatter plot analysis.

But scatter plots can also be performed on only a small group of genes collected in previous analyses, thereby enabling the user to focus on a relevant subset of the data. The second approach is for example useful if these genes are known or suspected to belong to one regulon, in which case they should show a similar behaviour under various conditions. Two-dimensional scatter plots provide an easy way to test this hypothesis, since currently marked genes can be labelled in the plot and thereby easily visualized (Fig. 4).

Images of the scatter-plots can be directly retrieved. For presentation or publication purposes, individual genes can be labelled with their names, or specific symbols can be assigned to groups of genes, in order to distinguish them.

4.3.2. *The cluster analysis (HeatMap) tool*

To perform more sophisticated expression analyses of multiple microarray datasets, the hierarchical clustering functions of the Cluster 3.0 Software (de Hoon *et al.*, 2004), an enhanced version of the Cluster Software[20], were integrated into REACT. This analysis assigns sets of genes into groups (the so-called clusters), so that the behaviour of the genes

---

[20] URL for the source code of the Cluster software: http://rana.lbl.gov/EisenSoftwareSource.htm

from within the same cluster is more similar to each other than to those in other clusters. Clustering is based on calculating a distance measure, which determines the similarity of two elements. During this calculation, the often n-dimensional data sets are reduced to their respective distances (one distance for each pair of objects). This less complex data set is then used as input for the final clustering. The corresponding algorithms achieving this differ significantly in their notion of what constitutes a cluster and in their efficiency of finding them.

The hierarchical cluster analyses embedded in the REACT suite provide a way to compare the expression behaviour of genes over multiple microarray datasets but also, if needed, to group and cluster arrays. The result is a two-dimensional, colour-coded matrix (or grid) in which each row represents one gene, while each column corresponds to one array dataset. Rows and/or columns are sorted according to their overall distances, and this clustering is illustrated by flanking distance trees, in which the length of the branches serves as a measure for similarity: The shorter the branches, the higher their similarity (Fig. 5).

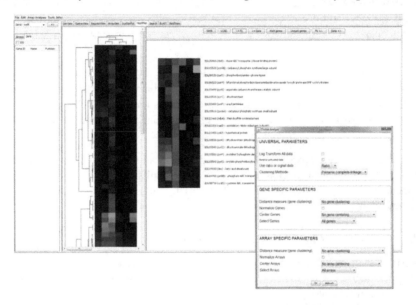

Figure 5. Example of a cluster analysis performed with the *HeatMap* tool. The inset on the right shows the parameter window for choosing the settings for a cluster analysis.

The complexity of the data is not lost, as all ratio or signal values for each gene within all arrays are visualized by the colour of the individual cells within the heat map grid. When ratio values are displayed, green colour indicates an increase (positive ratio value) and red a decrease (negative ratio value) of the expression in comparison to the control condition of the array, while the intensity of the colour is an indicator for the magnitude of change (Fig. 5). Signal values are coloured according to their percentage from the lowest and highest measured value within the array.

To run a cluster analysis, the user has first to decide, which genes and microarray datasets are to be included. Again, the active list of marked genes/arrays can directly be applied. Since REACT only serves as an interface to the Cluster 3.0 software package, its panel mimics the original input fields, with some modifications (inset to Fig. 5). The choice of parameters includes: clustering of only genes, arrays, or both; (ii) use of ratio or signal values; (iii) log-transformation of the data; (iv) removal of "untrusted" data (see above). Distance measures such as Euklidian distance, Kendall's tau, Pearson correlation or Spearman's rank correlation are available for both gene- and array-clustering. Moreover, genes and arrays can again be normalized as well as centered, as described for the scatter plot analysis above. For the final linkage, the user can choose between Pairwise Single, Pairwise Complete, Pairwise Centroide and Pairwise Average Linkage clustering methods[21].

The results of the cluster analysis are displayed in the *HeatMap* window (Fig. 5). This view is vertically split into two subpanels. The left panel displays the complete heat map, including the distance trees, while the right subpanel displays selected areas in more detail, including gene-IDs, names and descriptions. The heat map is interactive and selecting one row will directly open the corresponding *GeneView*. Marked genes are highlighted in red in the heat map.

To further analyse a certain gene cluster, it can directly be selected from the flanking distance trees, which are also interactive: Selecting any branch will mark the corresponding rows or columns. Intersections of selected rows and columns can be obtained and selected parts of the heat map can be displayed in higher resolution in the right subpanel of the *HeatMap* window, as described above.

The content of each subpanel can be exported both as an image file (different file formats can be chosen), as well as in tabulated form. Cluster results can also be stored and reloaded again, e.g. to enable the user to compare the clustering of specific groups of genes between different analyses.

4.3.3. The "Show regulons of marked genes" function

Co-expression – and therefore co-clustering – of groups of genes is a strong indication that they presumably belong to one regulon, i.e. are under the direct control of a common transcriptional regulator. In case of the two model bacteria currently implemented in the REACT suite, *B. subtilis* and *E. coli*, many of these regulons are already known.

To simplify the identification of known regulons within a marked group of genes derived from one of the above analyses, the "Show regulons of marked genes" function was implemented in the REACT suite, which displays all regulons to which at least one currently marked gene is associated in an additional window. Moreover, the results window will also list all operons and genes of any identified regulon, thereby providing a direct overview of the coverage of a given regulon within the group of marked genes identified by the cluster

---

[21] For details on clustering, see:
   http://bonsai.hgc.jp/~mdehoon/software/cluster/manual/Hierarchical.html#Hierarchical

analysis. As usual, the identifiers of the regulons, operons and genes function as internal links to the corresponding views, enabling a seamless integration with subsequent analyses of the identified transcriptional units.

This function therefore offers a very straightforward and easy-to-use approach to identify the regulators responsible for an observed co-expression of a group of genes.

4.4. Motif analysis tools

If the above mentioned function did not yield a direct insight into regulatory principles underlying an observed co-expression, the next step of a typical analysis would be to search for putative regulator binding sites in the upstream genomic regions of co-expressed genes and operons. To facilitate these analyses, the MEME/MAST tools from the MEME (Multiple EM for Motif Elicitation) suite[22] (Bailey et al., 2009) were incorporated into REACT. MEME allows the identification of short overrepresented sequence motifs in a group of unaligned sequences of different length. MAST is a sequence similarity search algorithm that utilizes motifs either provided by the user or from a previous MEME analysis, to search for similar motifs in genome sequences. Starting from the upstream regions of co-clustering genes, these two tools, if applied in combination, often allow to identify putative regulator binding site in novel regulons.

4.4.1. The MEME-Analysis tool

A prerequisite for any motif search is a collection of (upstream) sequences that are supposed to contain a common motif. In the REACT suite, this is facilitated by the "get upstream" function, which can be found in a number of views, including the Gene-, Operon- or MotifView. The latter also contains the panels for the MEME and MAST analyses. Again, REACT's motif discovery function is just an embedded interface to these freely available and well established tools, which are components of the MEME suite. MEME is a tool for discovering motifs in a group of related DNA or protein sequences. It represents motifs as position-specific scoring matrices (PSSM's), which describe the probability of each possible letter at each position within the gapless pattern. MEME uses statistical modelling techniques to automatically choose the best width, number of occurrences, and description for each motif to reduce the number of false-positive hits. Nevertheless, they can occur incidentally, especially if the motifs are very short, and therefore have to be validated experimentally, both in vivo and in vitro (Cao et al., 2002).

Like other analysis panels of REACT, the MEME view is also divided into two areas: in the upper part, the sequences and analysis parameters can be specified, while the results will be displayed in the lower panel (Fig. 6).

To start a MEME analysis, the user has to provide the sequences (in this case: upstream regions of genes), which are believed to share a common motif. This can be done by one of three ways: (i) Selection of upstream sequences from the "Upstream sequence" panel, (ii)

---

[22] URL for online access to the complete MEME suite at: http://meme.nbcr.net

directly pasting sequences into the respective sequence window of the MEME interface, or (iii) uploading an external file. The latter options enable the inclusion of sequences, which are not derived from the REACT database. Next, the number of allowed (or expected) motifs per sequence needs to be defined. Additional parameters include (i) the minimum and maximum motif-width, (ii) the maximum number of motifs to be discovered, (iii) a statistical threshold value, and (iv) limitation to palindromic sequences.

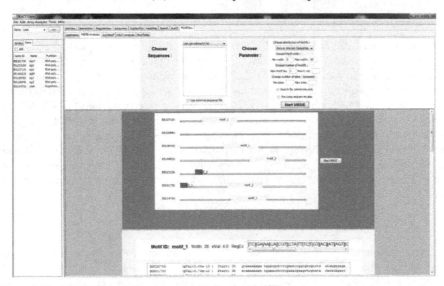

Figure 6. The MEME analysis interface embedded in the *MotifView*.

REACT`s MEME results consist of a graphical overview of the analysed sequences (Fig. 6) illustrating the occurrence and position of the motifs. Each motif is described by the following information: a motif ID, the length of the motif, a statistical value as a measure for the reliability of the motif, and a corresponding SequenceLogo as a graphical representation of the motif. As computable definitions, the description also includes the Regular Expression, an alignment of the motif from the analysed sequences, and the PSSM, which can all be exported. Alternatively, these definitions can be used directly for a MAST analysis to screen genome sequences from the REACT database for additional upstream regions containing this pattern (described in the following section) or stored in the REACT database for later analyses.

4.4.2. The MAST-Analysis tool

An important strategy to identify regulon members in large datasets, such as (multiple) genome sequences, is to screen them for the presence of sequence motifs, especially in intergenic regions, that are known or postulated to function as regulator binding sites. Such patterns can be derived from known operator sites described in other, closely related organisms (Wecke *et al.*, 2006), or from motifs identified by MEME analyses from collections of co-expressed loci, as described above. One way of testing predicted motifs *in silico* is to apply them to larger data sets. This not only allows the identification of additional putative matches, but it might also help to improve the motif through iterations. In the REACT suite, this can be done with the MAST analysis interface.

MAST is a tool for searching biological sequence databases for sequences that contain one or more copies of a known motif. The quality of a resulting hit is calculated as the strength of the similarity of the particular sequence to all motifs, based on statistical probabilities. MAST works by calculating match scores for each sequence in the database compared with each of the provided motifs. These initial scores are then converted into statistical probability values, which are used to determine the overall match of the sequence to the group of motifs. By this approach, the best fitting sequences in the analysed data set can ultimately be identified.

The MAST interface of the REACT suite is located within the *MotifView* and resembles the one for the MEME analysis both in appearance and overall logic. Two important parameters need to be defined by the user. The first one is the motif. It can be directly imported from a MEME result table, from the motifs stored in the REACT database (accessed via the motif table), but also manually imported from an external motif definition, expressed as a PSSM.

The second important parameter is the sequence database to be searched. REACT contains pre-compiled data files containing all upstream regions of the currently implemented two model organisms but also of all of the respective reference organism. These regions are defined as the 200 bases upstream of the start codon of each gene. Other parameters to be defined are the maximum number of sequences to be displayed, a probability threshold and if genes overlapping with the upstream regions should be displayed in the results.

After the analysis has been performed, a graphical overview of the results in the form of a block diagram is displayed. It shows the matching regions for each motif within each sequence, the direction of the match (forward or reverse), the gene ID to which the upstream region belongs, and a probability value indicating the match strength. The information can also be displayed as in tabulated form. As usual, the diagram is interactive and provides a direct link to the corresponding gene-specific information.

If additional promising matches could be identified, they can then be integrated into a new iteration of creating motifs with the MEME-tool and re-checking them with MAST. Again, the integrative nature of REACT will enable and simplify such follow-up analyses.

5. Operating the REACT suite

We will conclude this chapter with a brief summary of how the REACT suite can be navigated and modified. For this purpose, we will first describe a typical work flow through the features of REACT from the perspective of a user (5.1). In the second section, we will specifically address the rights and options of REACT-administrators (5.2). Finally, we will provide a brief summary of the REACT concept and infrastructure (5.3).

5.1. Navigating REACT: The user approach

The functionality of the REACT suite relies on curated and comprehensive data that is provided by the organism-specific REACT database. It provides three different types of data: (i) gene-centric data (derived from genome sequences and their annotation), (ii) array-centric data (extracted from microarray databases and individual sources of transcriptome experiments), and (iii) motif data (based on experimental and computational evidence).

While there are many ways to use the REACT suite, it was developed with the goal in mind to enable the user to identify and characterize regulons starting from in-depth analyses of microarray datasets. Here, we will illustrate a typical workflow through the REACT suite (Fig. 7), in order to highlight the concept of REACT by connecting the central features that have so far been primarily described in isolation in the previous sections.

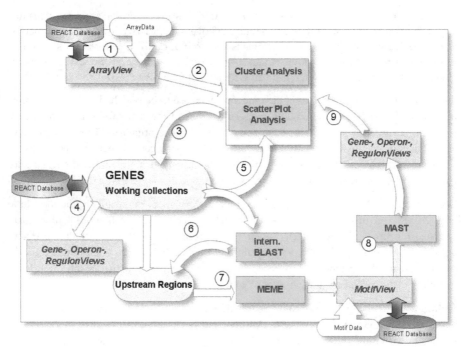

Figure 7. Work flow through the REACT suite. Major views are indicated by light grey, analysis tools by the green colour. The numbers refer to the description in the text.

A typical experiment could start with importing new microarray datasets ① to be subsequently analysed in detail by scatter plot or cluster analysis ②. These initial studies will presumably be performed genome-wide, but with a limited number of relevant microarray datasets. As a result, groups of interesting genes will be identified ③ that respond in a condition-specific manner and could potentially be co-expressed and therefore co-regulated. All unknown genes can be subjected to in-depth analyses, primarily using the information stored in the *GeneView* (including the external links), but to some extend also data from the *OperonView*, and *RegulonView* ④. Moreover, the group of selected genes could also be subjected to a second round of Cluster analysis, now incorporating a more diverse set of array conditions on this limited number of genes, to refine the clustering ⑤. Genes of interest derived from any of these studies can be selected and thereby added to the list of marked genes. For genes of interest that cannot be associated with known regulons, the upstream regions will be retrieved for the subsequent steps of the regulon annotation. To increase the chance of identifying regulator binding sites, internal BLAST analyses can be performed to extract upstream regions from orthologous genes from closely related species ⑥, assuming that they are subject to the same regulation. The collection of upstream regions will then be subjected to a MEME analysis to identify common sequence motifs as candidates for putative regulator binding sites ⑦. The motif definitions will be incorporated into the *MotifView* and subsequently used to screen genome sequences for additional candidates, using the MAST tool ⑧. If new candidate target genes preceded by the conserved motif could be identified, they will then be selected and subjected to the comprehensive studies described above, including Cluster analysis to compare their behaviour to the group of genes initially selected ⑨.

Enabling such iterative and interactive processes that rely on both sequence-based and array-based data and analysis tools is a major advantage of the REACT suite. Because of its concept and architecture, the necessary information and data flow can be controlled easily and the analyses can be performed efficiently.

5.2. Modifying REACT: The administrator mode

In the age of omics, new genome sequences and microarray studies are published with ever-increasing speed. It is therefore important that a REACT database, once established, can be updated regularly to grow with the increase of available data and information. But as a precautious measure to avoid data corruption and thereby ensuring the integrity of the database, it is advisable that not all users have the right to modify the core data at all times during analyses. REACT has therefore implemented two different user roles: the REACT-user normally works in the "read-only" mode. This will allow him to browse the data, perform analyses, and export data to external files. In contrast, login as a REACT-administrator enables the user to permanently import additional data (such as microarray datasets or new reference genomes), to edit data already implemented in the REACT database, and even to change the main views of REACT by incorporating additional links and features.

When logged in as REACT-administrator, most data displayed in the different views can be edited manually. To prevent unintentional data corruption, data can only be changed after deliberately switching into the edit-mode via the appropriate buttons, provided in each view. In the edit-mode, all editable data is displayed in green and all links are disabled. Any changes applied to the data remains transient until they are confirmed by the REACT-administrator and thereby sent to the REACT-server and stored permanently.

However, some data fields are not editable, as REACT uses them as immutable internal references (e.g. as primary and secondary database keys) to identify the complete dataset. This includes the names and IDs of genes, operons, or microarray datasets, as well as DNA and amino acid sequences from the *GeneView*, which are derived from and defined by the respective genomic sequence.

A REACT-administrator can also define new data fields for the above listed views according to the individual requirements, including plain text fields and numeric fields. Moreover, new external links can also be added to the views. While it is quite easy to generate plain text or numeric fields (as just the field name and type have to be defined within the REACT-administrator dialogue), creation of additional link-fields is technically a bit more demanding.

In addition to the aforementioned options, REACT-administrators can import additional array data, create new array sets or change the assignments of arrays to a set. They can also store motifs computed during a MEME analysis permanently within the REACT database or define new regulons. In the *RegulonView*, new operons can be connected to or removed from the regulons.

5.3. Expanding REACT: Embedding new organism-specific databases

REACT was initially developed for the analysis of two model organisms, *E. coli* and *B. subtilis*. But given the wealth of knowledge already available for these organisms, the potential of REACT may be even higher when applied to genomic and expression data of other, less well-characterized organisms.

Therefore, REACT is equipped with a small set of additional tools that enables researchers with little knowledge of programming languages or database administration to create new organism-specific REACT databases from scratch. Following the instructions provided by the software, the user has to download freely available files from sources like the NCBI, Uniprot or MicrobesOnline databases that contain the data used by REACT. Additional information (e.g. links to PFAM or PDB) will be obtained from the KEGG web service via SOAP/WSDL, again without the need for more than very basic user interaction.

After the creation of an initial, empty REACT database (done by importing a provided sql-file into the SQL database), the information contained in the downloaded files and provided by KEGG are parsed by helper tools provided by the REACT package, again minimizing user interaction.

Users with basic programming knowledge will then be able to extend the new REACT database by parsing data from additional data sources, depending on the organism chosen and the focus of the respective database. Subsequently, additional data needed by REACT (e.g. interbl BLAST databases) will be computed automatically. The user will now be able to connect with this newly created REACT database, in order to upload the first array sets.

5.4. Developing REACT: Concept, sources and infrastructure

TAs mentioned previously, the major aim of the REACT bioinformatics toolkit was the creation of an intuitive and interactive graphical user interface that allows an integrative view on genomic and microarray data and provides combined access to various bioinformatics tools commonly used in comparative genomic and transcriptomic studies. The overall structure of the REACT suite is illustrated in Fig. 8.

In the current release, the tools listed in Table 1 are integrated into the REACT suite. The software was implemented using a client/server architecture, enabling the parallel and locally distributed work of one to multiple users (Fig. 8). The REACT-server is the central computer running the database-managing software (MySQL), as well as all internal and integrated third-party analysis tools. The users will solely work with the corresponding client program, which can be installed on the personal computers or laptops of all users. Client and server are communicating via intra- or internet using remote method invocation (RMI) techniques. REACT is implemented as a java swing application, therefore client and server should run under a variety of operating systems depending only on the Java Runtime Environment (Version 5 or higher). However, in case of the server, this is limited by the external tools, as some of them (e.g. the MEME suite) depend on a Linux / Unix environment. To circumvent this limitation, REACT was developed and tested for being executable on Windows OS using Cygwin (1.5.x or higher), which is a Linux emulator for Windows and provides substantial Linux API functionality.

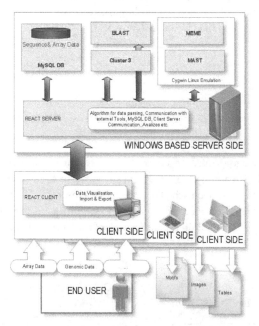

Figure 8. Structure, components and data flow of the REACT suite. See text for details.

Name	Version	Link	Reference
Blast	2.2.x	ftp://ftp.ncbi.nlm.nih.gov/blast/executables/release/LATEST/	(Altschul *et al.*, 1990)
Cluster 3	3.0	http://bonsai.hgc.jp/~mdehoon/software/cluster/	(de Hoon *et al.*, 2004)
MEME suite	4.0.0	http://meme.sdsc.edu/meme/meme-download.html	(Bailey *et al.*, 2009)
Cygwin	1.7.5	http://www.cygwin.com/install.html	n.a.
MySQL	5.5.x	http://www.mysql.de/downloads/mysql/	n.a.

Table 1. Third party software tools implemented in the REACT suite. "n.a.", not applicable.

6. Conclusion

This chapter aimed at providing a thorough overview of the concept and functions of the REACT suite, a bioinformatics toolkit that was developed to simplify regulon predictions and comparative transcriptomic analyses for biologists with little to no background in bioinformatics. REACT was written in the believe that it will provide a powerful, yet simple-to-use platform that will hopefully also support the work of other research groups in extracting meaningful data from transcriptome studies with the help of comparative genomics. The complete REACT suite, including the databases for *B. subtilis* and *E. coli*, are available from the corresponding author upon request.

Author details

Peter Ricke and Thorsten Mascher[*]
Ludwig-Maximilians-University Munich, Germany

Acknowledgement

The authors would like to thank Tina Wecke for beta-testing of the REACT suite, providing the figures and critical reading of the manuscript. Work in the Mascher lab is financially supported by grants from the Deutsche Forschungsgemeinschaft (DFG). Development of the REACT suite was enabled by funding from the 'Concept for the future' of the Karlsruhe Institute of Technology (KIT) within the framework of the German Excellence Initiative.

7. References

Altschul, S. F., W. Gish, W. Miller, E. W. Myers & D. J. Lipman, (1990) Basic local alignment search tool. *J Mol Biol* 215: 403-410.

---

[*] Corresponding Author

Bailey, T. L., M. Boden, F. A. Buske, M. Frith, C. E. Grant, L. Clementi, J. Ren, W. W. Li & W. S. Noble, (2009) MEME SUITE: tools for motif discovery and searching. *Nucleic acids research* 37: W202-208.

Bairoch, A., (2000) The ENZYME database in 2000. *Nucleic acids research* 28: 304-305.

Barrett, T., D. B. Troup, S. E. Wilhite, P. Ledoux, C. Evangelista, I. F. Kim, M. Tomashevsky, K. A. Marshall, K. H. Phillippy, P. M. Sherman, R. N. Muertter, M. Holko, O. Ayanbule, A. Yefanov & A. Soboleva, (2011) NCBI GEO: archive for functional genomics data sets - 10 years on. *Nucleic Acids Res* 39: D1005-1010.

Brazma, A., (2009) Minimum Information About a Microarray Experiment (MIAME)--successes, failures, challenges. *TheScientificWorldJournal* 9: 420-423.

Brazma, A., P. Hingamp, J. Quackenbush, G. Sherlock, P. Spellman, C. Stoeckert, J. Aach, W. Ansorge, C. A. Ball, H. C. Causton, T. Gaasterland, P. Glenisson, F. C. Holstege, I. F. Kim, V. Markowitz, J. C. Matese, H. Parkinson, A. Robinson, U. Sarkans, S. Schulze-Kremer, J. Stewart, R. Taylor, J. Vilo & M. Vingron, (2001) Minimum information about a microarray experiment (MIAME)-toward standards for microarray data. *Nature Genetics* 29: 365-371.

Cao, M., P. A. Kobel, M. M. Morshedi, M. F. Wu, C. Paddon & J. D. Helmann, (2002) Defining the *Bacillus subtilis* s^W regulon: a comparative analysis of promoter consensus search, run-off transcription/macroarray analysis (ROMA), and transcriptional profiling approaches. *J Mol Biol* 316: 443-457.

de Hoon, M. J. L., S. Imoto, J. Nolan & S. Miyano, (2004) Open source clustering software. *Bioinformatics* 20: 1453-1454.

Dehal, P. S., M. P. Joachimiak, M. N. Price, J. T. Bates, J. K. Baumohl, D. Chivian, G. D. Friedland, K. H. Huang, K. Keller, P. S. Novichkov, I. L. Dubchak, E. J. Alm & A. P. Arkin, (2010) MicrobesOnline: an integrated portal for comparative and functional genomics. *Nucleic acids research* 38: D396-400.

Demeter, J., C. Beauheim, J. Gollub, T. Hernandez-Boussard, H. Jin, D. Maier, J. C. Matese, M. Nitzberg, F. Wymore, Z. K. Zachariah, P. O. Brown, G. Sherlock & C. A. Ball, (2007) The Stanford Microarray Database: implementation of new analysis tools and open source release of software. *Nucleic Acids Res* 35: D766-770.

Finn, R. D., J. Mistry, J. Tate, P. Coggill, A. Heger, J. E. Pollington, O. L. Gavin, P. Gunasekaran, G. Ceric, K. Forslund, L. Holm, E. L. Sonnhammer, S. R. Eddy & A. Bateman, (2010) The Pfam protein families database. *Nucleic Acids Res* 38: D211-222.

Gama-Castro, S., H. Salgado, M. Peralta-Gil, A. Santos-Zavaleta, L. Muniz-Rascado, H. Solano-Lira, V. Jimenez-Jacinto, V. Weiss, J. S. Garcia-Sotelo, A. Lopez-Fuentes, L. Porron-Sotelo, S. Alquicira-Hernandez, A. Medina-Rivera, I. Martinez-Flores, K. Alquicira-Hernandez, R. Martinez-Adame, C. Bonavides-Martinez, J. Miranda-Rios, A. M. Huerta, A. Mendoza-Vargas, L. Collado-Torres, B. Taboada, L. Vega-Alvarado, M. Olvera, L. Olvera, R. Grande, E. Morett & J. Collado-Vides, (2011) RegulonDB version 7.0: transcriptional regulation of *Escherichia coli* K-12 integrated within genetic sensory response units (Gensor Units). *Nucleic Acids Res* 39: D98-105.

Lechat, P., L. Hummel, S. Rousseau & I. Moszer, (2008) GenoList: an integrated environment for comparative analysis of microbial genomes. *Nucleic Acids Res* 36: D469-474.

Letunic, I., T. Doerks & P. Bork, (2009) SMART 6: recent updates and new developments. *Nucleic Acids Res* 37: D229-232.

Liolios, K., I. M. Chen, K. Mavromatis, N. Tavernarakis, P. Hugenholtz, V. M. Markowitz & N. C. Kyrpides, (2010) The Genomes On Line Database (GOLD) in 2009: status of genomic and metagenomic projects and their associated metadata. *Nucleic Acids Res* 38: D346-354.

Rose, P. W., B. Beran, C. Bi, W. F. Bluhm, D. Dimitropoulos, D. S. Goodsell, A. Prlic, M. Quesada, G. B. Quinn, J. D. Westbrook, J. Young, B. Yukich, C. Zardecki, H. M. Berman & P. E. Bourne, (2011) The RCSB Protein Data Bank: redesigned web site and web services. *Nucleic Acids Res* 39: D392-401.

Sierro, N., Y. Makita, M. de Hoon & K. Nakai, (2008) DBTBS: a database of transcriptional regulation in *Bacillus subtilis* containing upstream intergenic conservation information. *Nucleic Acids Res* 36: D93-96.

Sigrist, C. J., L. Cerutti, E. de Castro, P. S. Langendijk-Genevaux, V. Bulliard, A. Bairoch & N. Hulo, (2010) PROSITE, a protein domain database for functional characterization and annotation. *Nucleic Acids Res* 38: D161-166.

Tatusov, R. L., E. V. Koonin & D. J. Lipman, (1997) A genomic perspective on protein families. *Science* 278: 631-637.

Utts, J. M., (2005) *Seeing through statistics*. Thomson Brooks.

Wecke, T., B. Veith, A. Ehrenreich & T. Mascher, (2006) Cell envelope stress response in *Bacillus licheniformis*: Integrating comparative genomics, transcriptional profiling, and regulon mining to decipher a complex regulatory network. *J. Bacteriol.* 188: 7500-7511.

Chapter 5

Genome-Wide RNAi Screen for the Discovery of Gene Function, Novel Therapeutical Targets and Agricultural Applications

Hua Bai

Additional information is available at the end of the chapter

1. Introduction

The phenomenon of double-stranded RNAs (dsRNAs)-mediated gene silencing or RNA interference (RNAi] was first discovered in nematode *Caenorhabditis elegans* by Andrew Fire and Craig Mello in 1998 [1]. This great discovery gives rise to a fast-growing field and leads to the identification of novel RNAi pathways by which small interference RNAs (siRNAs) regulate gene expression and gene functions. Collective evidence suggests that the RNAi pathway is conserved in many eukaryotes and this pathway can be triggered by either exogenous or endogenous small interference RNAs. Exogenous dsRNAs (e.g. a virus with an RNA genome) are typically required a membrane transporter for dsRNA uptake into the cytoplasm, while endogenous small interference RNAs (e.g. microRNAs) are encoded in the genome. The precursors of both dsRNA and microRNA are first cleaved into short interference RNAs by a ribonuclease III (RNaseIII) enzyme, Dicer. Then these short interference RNAs initiate RNAi process when interacting argonaute proteins in the RNA-induced silencing complex (RISC). The small interference RNAs normally consist of 20~30 nucleotides. They can repress the transcription of message RNAs containing homologous sequences by either post-transcriptional gene silencing (PTGS) or transcriptional gene silencing (TGS) [2].

In their *Nature* paper, Fire and Mello wrote: 'Whatever their target, the mechanisms underlying RNA interference probably exist for a biological purpose'. Indeed, it's been shown that there are numerous cellular and physiological functions linked to RNAi [1]. For example, viral defense has been proposed to be the primary function of RNAi in both plants and flies [3]. In plants, virus infection could trigger sequence-specific gene silencing [4].

Plant RNAi forms the basis of virus induced gene silencing (VIGS), proofed from the genetic links between virulence and RNAi pathways [5-6]. On the other hand, endogenous microRNAs (about 1000 microRNAs in human genome) [7] play essential roles in controlling cellular functions. For example, the early discovered microRNAs, such as *lin-4* and *let-7* of *C. elegans*, were identified to regulate developmental timing [8-9]. Following the identification of *let-7* in *C. elegans* and later in fruit flies *Drosophila melanogaster* (hereafter I will refer it as '*Drosophila*'), it is soon realized that *let-7* belongs to a conserved microRNA family in many species. Besides the regulation on development, many microRNAs have been found to control key physiological processes, such as lipid metabolism [10] and insulin sensitivity [11]. The dysregulation of microRNA may result in many human diseases. A mutation in the seed region of miR-96 causes hereditary progressive hearing loss [12]. Some microRNAs have also been linked to cancer [13].

Soon after the discovery of dsRNA-mediated RNAi in 1998, gene silencing through RNAi was quickly developed as a powerful tool or technique in functional genomics studies. Comparing to forward genetics tools (e.g. EMS-induced mutagenesis screens), RNAi is one of effective reverse genetic tools, especially for non-model organisms and mammalian systems in which genetics is difficult. The advantage of applying RNAi in function studies becomes even more apparent when the whole genome sequences of model organisms (e.g. *C. elegans* and *Drosophila*) were completed in the late 90's and early 2000's [14-15]. In the post-genome era, utilizing high-throughput platforms and innovative bioinformatics tools, many large-scale RNAi screens has been successfully applied for the discovery of novel gene function associated with many important aspects of biology such as signal transduction [16], cell proliferation [17], metabolism [18], host-pathogen interactions [19] and aging [20-21]. Through these genome-wide RNAi screens, we have gained new insights on novel players in many key biological processes and complexity of cellular signaling networks. Cell-based and *in vivo* RNAi screen has been extensively reviewed in the past [22-23].In this book chapter, I will focus on the recent development of high-throughput RNAi screen for functional analysis in cultured cells and *in vivo* systems, as well as its applications on functional genomics and the discovery of novel therapeutic drug and agricultural targets.

2. RNAi screen methods

Despite the challenges from off-target effects and false discovery during the data analysis, genome-wide RNAi screens have benefitted from improved RNAi delivery methods, automated high-content image system and robust statistic analysis [23]. In the following section, I will compare the different reagent delivery methods, read-out assays, and off-target effects in various systems and platforms. I will also provide several examples from recent studies using genome-wide RNAi screen in cultured cells.

2.1. Cell-based RNAi screen

Genome-wide RNAi screen in cultured cells or primary cells provides an opportunity to systematically interrogate gene function. Now large-scale RNAi screens have been routinely performed in *Drosophila* and mammalian cultured cells, as well as in primary cells. RNAi screen with *Drosophila* and mammalian cells has already led to important discovery in a wide variety of topics, including signal transduction, metabolism and cancer [22]. In general, a cell-based RNAi screen involves four major steps: [1]. RNAi library selection; [2]. Incubation of appropriate cell lines with RNAi reagents that are pooled or individually arrayed into 96- or 384-well plates; [3]. After additional treatments (if applicable), cells are subjected to the automated plate reader to quantify the specific readout (e.g. changes in cell morphology or fluorescence and luminescence signals from study targets); [4]. High-content image data analysis.

RNAi library and reagent delivery methods. Since the completion of *C. elegans* genome sequencing in 1998 [14], more and more eukaryotic genomes have been sequenced, which makes it possible to produce whole-genome RNAi libraries for functional genomics studies. Typically, long dsRNAs are used for RNAi screen in *Drosophila* cells, while synthetic siRNAs or vector-based short hairpin RNAs (shRNAs) are commonly used for mammalian cells [24]. Several *Drosophila* cell lines (e.g. S2 and Kc167) can directly take up dsRNA without the help of transfection reagents, which provides great advantages in high-throughput RNAi screens [25-26]. For mammalian cells, RNAi reagents are transient transfected into the cells. Therefore, not only the variation of transfection efficiency among cell lines and experimental replicates will affect the RNAi knockdown effects, but also the doubling time will affect the duration of gene silencing. Compared to synthetic siRNA method, vector-based shRNA technology combining with viral delivery methods provides robust gene silencing for a longer period of time. Besides, vector-based shRNA approaches make it possible to build renewable and cost-effective RNAi libraries.

Now RNAi libraries are available for many model organisms, especially *Drosophila* and mammalian cells (Table. 1). These RNAi libraries normally contain the collection of RNAi reagents (dsRNAs, siRNAs and shRNAs) for all annotated genes in the genome. In the RNAi libraries, typically there are several different dsRNAs or siRNAs corresponding to each gene. For example, The GeneNet™ Human 50K siRNA library from System Biosciences contains 200,000 siRNA templates targeted to 47,400 human transcripts (~4 different siRNA sequences per transcript) (http://www.systembio.com/rnai-libraries). In *Drosophila* DRSC 2.0 RNAi collection, there are 1-2 dsRNAs per gene, including genes that encode proteins and non-coding RNAs (http://www.flyrnai.org). Beside genome-wide libraries, many pathway libraries or sub-libraries are also available for silencing specific signal pathways or multi-gene families (e.g. kinases & phosphatases library, G protein-coupled receptor library, apoptosis & cell cycle library, etc.).

Read-out assays. Together with luminescent and fluorescent reporter-based image analysis and improved high-content image processing packages (e.g. CellProfiler [27] , various read-outs are used in RNAi screens, including the expression changes in target genes or proteins, post-translation modification, metabolic processes, and changes in sub-cellular localization patterns. Most of these read-outs are based on the measurement of the intensity of luminescent and fluorescent reporters. Although RNAi screens for cell morphology (e.g. cytoskeletal organization and simple cell shape) have been previously reported [28-30], complex cell shape and structure-based read-outs still remain problematic. Frequently, additional treatments are performed before the read-out assays in RNAi screens. These treatments can be drugs, pathogens, or stress inducers [31-33].

Name	Species	Type	Link
Genome-wide RNAi libraries			
Drosophila RNAi Screen Center	Fruit fly	dsRNA	www.flyrnai.org/
DKFZ Genome RNAi	Fruit fly	dsRNA	www.genomernai.org
Open Biosystems	Human, mouse	siRNA , shRNA	www.openbiosystems.com/RNAi
Sigma	Human, mouse	siRNA , shRNA	www.sigmaaldrich.com
SBI	Human, mouse	siRNA , shRNA	http://www.systembio.com/rnai-libraries
Pre-defined or custom RNAi libraries			
Ambion	Human, mouse	siRNA , shRNA	www.invitrogen.com/sirna
Qiagen	Human, mouse	siRNA , shRNA	www.qiagen.com
Dharmacon	Human, mouse	siRNA , shRNA	www.dharmacon.com
The Netherlands Cancer Institute (NKI)	Human	shRNA	www.lifesciences.sourcebioscience.com/

Table 1. List of RNAi libraries used in cell cultures

Off-target effects. False positive or false negative results are commonly associated with high-throughput studies, including genome-wide RNAi screens [34]. The false discovery in RNAi screens can be caused by instrument errors, statistical noises, low knock-down efficiency and off-target effects of RNAi reagents. Off-target effects of RNAi reagents usually include: [1]. A general interference on endogenous RNAi pathway; or [2] Sequence-dependent effects on the expression of non-target genes. In order to minimize off-target effects, it is suggested to perform sufficient replication experiments and choose two or more RNAi reagents that target different regions of the coding sequences. Sequence-dependent off-target effects can be avoided by selecting sequences that do not contain 19 or more base pairs of contiguous nucleotide identity to other genes in the genome [34-35].

Recent cell-based RNAi screen studies. Genome-wide RNAi screens have been primarily conducted in both *Drosophila* and mammalian cultured cells (Reviewed in [22]. These screens are involved in studies on a variety of biology processes, such as signal transduction, metabolism, cancer, stem cells. The cell-based RNAi screens have yielded tremendous amount of novel discoveries and greatly promoted our understanding on many basic biological processes, molecular functions and complexity of cellular networks. New findings from genome-wide RNAi screens have led to the identification of novel components of canonical signaling transduction pathways, such new players of ERK pathway [16] and phosphorylation networks regulating JUN NH2-terminal kinase (JNK) pathway [36]; the role of *S1pr2* gene (Sphingosine-1-phosphate receptor 2) in insulin signaling [37]; novel modulators of p53 pathway [38]; and key genes that are essential for the proliferation of cancer cells [17]. Recently, an integrative approach with RNAi screen and whole genome structural analysis identified IKBKE kinase as a breast cancer oncogene [39]. Beside its application in studying signaling pathways, cell-based RNAi screens are also performed to understand the cellular responses to pathogens. For example, recent RNAi screens identified novel host factors that are required for dengue virus propagation [40] and influenza virus replication [41]. I will discuss more detail on some of genome-wide RNAi screens in section. 3.

2.2. *In vivo* RNAi screen

Many complex phenotypes, such as aging, cannot be tested in a cell-based assay, thus *in vivo* functional analysis is required. *In vivo* RNAi screen is one of such approaches to study gene function at an organism level. *C. elegans* and *Drosophila* are two model organisms that are commonly used in *in vivo* RNAi screens. Although *ex vivo* RNAi screens have been done by introducing shRNA-transfected cells into mice [42-43], direct *in vivo* RNAi screen in mice is still under development. Most importantly, *in vivo* RNAi makes it possible for gene function studies in species lacking classical genetic tools, even including species without whole genome sequences (Usually next-generation sequencing is used to identify gene coding sequences and to facilitate RNAi reagent design and production).

In vivo genome-wide RNAi screen was first reported in *C. elegans* [44]. In *C. elegans*, dsRNA-mediated RNAi effects are systemic and heritable [1], although gene knockdown is less efficient in the nervous system than in other tissues. *In vivo* genome-wide RNAi screens have been performed in the studies of aging [20-21, 45-46], metabolism [47] and microRNA pathways in *C. elegans* [48]. It is relatively easy to deliver dsRNA into *C. elegans*. Typically, dsRNAs are introduced to worms by soaking the animals in dsRNA solution [49], by injection of dsRNA [1], or by feeding the animals dsRNA-expressing bacteria [44]. The last method is commonly used to generate genome-wide RNAi libraries. These *C. elegans* RNAi libraries are now available from Ahringer lab RNAi collection and Open Biosystems (Table. 2).

In contrast, dsRNA feeding does not appear to work for gene silencing in *Drosophila*, while RNAi via injection of dsRNA is effective only in certain embryonic stages. Therefore, transgenic RNAi approach has been developed to express a double-stranded 'hairpin' RNA from a transgene. In *Drosophila*, RNAi is cell-autonomous, so that gene silencing can be easily performed in tissue- and spatial-specific manner by using the binary GAL4/UAS expression system. Currently, there are three groups that have generated genome-wide transgenic RNAi *Drosophila* stains [50] (Table. 2). These transgenic RNAi stains are all expressing inverted-repeat hairpin RNAs once crossing to appropriate Gal4 drivers. Recently, it's been shown that small hairpin RNAs (~19 nt) can trigger stronger gene inactivation than long hairpin RNAs. Therefore new constructs expressing small hairpin RNAs were generated to produce second generation of transgenic RNAi *Drosophila* stains at Harvard medical school (Table. 2). In the past several years, a number of genome-wide RNAi screens in *Drosophila* have been conducted to study the major signaling pathways [51], as well as many important disease models [52-55].

For example, a genome-wide obesity gene screen revealed hedgehog signaling as one of major adipose tissue regulators [18], while genome-wide Parkinson's disease modifier screen identified novel *Park* and/or *Pink1*-interacting genes [56]. In the following section, I will discuss some of these RNAi screens in more detail.

Name	Species	Type	Link
Ahringer lab RNAi Library at Geneservice	Nematode	Bacterial clone	www.lifesciences.sourcebioscience.com/
Open Biosystems	Nematode	Bacterial clone	www.openbiosystems.com/RNAi/
Transgenic RNAi project at Harvard Medical School	Fruit fly	Long dsRNA, Short shRNA	www.flyrnai.org/TRiP-HOME.html
Vienna Drosophila RNAi Center	Fruit fly	Long dsRNA	stockcenter.vdrc.at
NIG-FLY	Fruit fly	Long dsRNA	http://www.shigen.nig.ac.jp/fly/nigfly/

Table 2. List of RNAi libraries used for *in vivo* systems

2.3. Advantage and limitation of cell-based and *In vivo* RNAi screens

Unlike forward genetic screens where mutations are randomly generated, RNAi screens provide a fast way to link phenotypes of interest to a precise gene. Beside, RNAi screens are generally performed in a genome-wide scale which brings us a comprehensive view of gene functions. Both cell-based and *in vivo* RNAi screens are highly effective and less labor-intensive on the discovery of gene functions when compared to traditional mutagenesis screens. Furthermore, cell-based and *in vivo* RNAi can be applied to study gene functions in species lacking classical genetics tools.

In the past decade, genome-scale *in vitro* RNAi screens have been successfully applied for gene discovery and understanding fundamental biological processes and cellular signal pathways. A variety of RNAi libraries for cell-based RNAi screens have been developed for both invertebrate and vertebrate systems. Nowadays cell-based RNAi screens are relatively less expensive, and have become a fast and user-friendly platform for functional genomics studies. On the other hand, complex phenotypes that cannot be analyzed in cell-based RNAi screens, are normally directly studied *in vivo*. When compared to forward genetic screens where mutations are occurred in every cell and many mutations lead to developmental defect or lethality, *in vivo* RNAi screens can be performed in various developmental stages and different tissues. This is especially useful when adult-specific functions of target genes are studied and these genes are essential for the development. Currently, most of *in vivo* RNAi screens are conducted in *C. elegans* and *Drosophila* due to the availability of a tremendous amount of resources and advanced genetic tools. In contrast, *in vivo* RNAi screen in mice is still in early stage.

3. Application of RNAi screen

Genome-wide RNAi screen has greatly advanced our understanding on many fundamental biology problems, from signaling transduction pathways to complex phenotypes. Furthermore, the results from RNAi screen can be used to design future theroputic drugs and crop protection reagents. In the following section, I will discuss several applications using RNAi screen approaches.

3.1. Deciphering cellular signaling pathways and complex traits

RNAi is one of the most powerful tools in functional genomics studies. Genome-wide RNAi screens have accelerated our understanding of basic biological functions and cellular signal networks, as well as the novel modulators of diseases. Follow-up experiments are usually performed to confirm the screen results and further study the underlying molecular mechanisms of identified genes or pathways. *In vivo* RNAi screens have also been conducted to study complex traits, such as aging. *C. elegans* is the primary model organism used in longevity gene screen, not only because high-throughput RNAi experiments are relatively easy to do in *C. elegans*, but also because of its short lifespan [57].

3.1.1. Understanding signaling pathways

Our understanding on canonical signal pathways is rapidly evolving and many new components or modulators are being identified with the help from improved technologies, including genome-wide RNAi screen. In the past decade, RNAi screens have been applied for deciphering many classical signal pathways, such as Notch, Wnt, and ERK signalings. Receptor tyrosine kinases (RTKs) are probably one of most critical protein families that regulate development, cell proliferation and growth. One of RTK families, insulin signaling plays important roles in controlling metabolism and growth. Disrupted insulin signaling leads to many human diseases, such as diabetes. To facilitate the underlying mechanism of diabetes and identify novel components and modulators of the insulin signaling pathway, a RNAi screen was conducted using 3T3-L1 adipocytes [37]. About 313 obesity and diabetes related genes were selected in the RNAi screen. The release of free fatty acid (FFA) was used as a read-out, since insulin-dependent FFA release is an indicator of insulin resistance. This screen showed that RNAi against 126 candidate genes resulted in significant changes of FFA release. After future filtering, *S1pr2* gene was identified as one of key regulators of insulin signaling. Increased plasma insulin levels were detected in male *S1pr2* -/- knockout mice, suggesting there is a potential link between *S1pr2* and insulin resistance [37].

One of RTK downstream effectors is ERK signaling pathways. Misregulated RTK/ERK signaling leads to developmental disorder and many human diseases (e.g. cancer). A recent RNAi screen study using *Drosophila* cell lines has identified 331 regulators of ERK pathway, suggesting a number of integrated signal pathways in the regulation of fine-tuned ERK signaling [16]. In this study, fluorescently-conjugated phospho-ERK antibodies were used to monitor the changes of phosphorylated ERK upon insulin stimulation, which is the first time that a RNAi screen uses post-translation modification as a read-out assay.

3.1.2. Identification of longevity genes

Aging is one of the complex traits that are controlled by a large number of genes or the interaction between multiple genes/pathways. The first longevity pathway, insulin/ IGF-1 pathway, was identified in C. *elegans* in early 90's [58]. Following studies have shown that TOR (Target of Rapamycin) [59] and AMP kinase signaling [60] are also involved in longevity regulation. To explore other potential longevity assurance genes/pathways, two systematic RNAi screens for longevity genes were independently conducted at almost the same time in C. *elegans*. [45-46]. Both groups used the Ahringer bacterial RNAi libraries, although they chose different worm strains in the RNAi screen. Initially, the maximum lifespan of each RNAi clone was monitored, due to the tremendous work on the large-scale lifespan screen. In one screen, 89 genes were identified to be involved in lifespan regulation. These candidate genes encode diverse biological functions, including metabolism, mitochondrial functions, signal transduction, protein turnover, and so on. In contrast, 29 genes were identified in another screen. Although both groups are able to re-discover the genes in insulin/IGF-1 signaling, only three newly identified genes are shared by these two screens. This may be due to the different worm stains used in these two screens, plus high level of false positive/negative hits and off-target effects. Knockdown of these three genes led to robust lifespan extension [57].

Although genome-wide RNAi screens for longevity genes have not been reported in other species, large-scale genetic screens were performed recently. A screen of 564 single gene deletion strains was conducted to identify longevity genes in budding yeast [59]. Deletion of 10 genes led to extended replicative lifespan. Among them, many genes are involved in TOR pathway, suggesting a link between nutrient sensing and longevity. Recently, a large-scale misexpression screen for *Drosophila* longevity genes was reported. In this screen, a total of 15 longevity genes were identified, including genes involved in autophagy, mRNA synthesis, intracellular vesicle trafficking and neuroendocrine regulation [61]. With more longevity gene screens from other species, a cross-species comparison of these large-scale screens may provide us a list of conserved genes/pathways in regulating longevity.

3.2. Identification of therapeutic drug targets

A number of genome-wide RNAi screens have led to the identification of novel modulators of human diseases. RNAi screens have become an effective tool to identify and validate drug targets and to enhance novel drug discovery. On the other hand, RNAi-based therapies have been developed to target viral infection, cancer, cardiovascular disease and neurodegenerative diseases using specific shRNAs, although we should always keep in mind that there are some drawbacks and concerns of this technology, such as off-target effects, activation of endogenous RNAi pathways, individual genetic variation.

Traditional cancer drug discoveries still focus on a handful of known oncogenes. It remains a key challenge in identifying new targets. Application of genome-wide RNAi screens in novel target discovery greatly enhanced cancer drug discovery. In the past few years, several RNAi screens were performed using cancer cell lines to explore the essential genes that are required for survival and proliferation of cancer cells. One of these studies identified more than 250 genes are essential for the proliferation of cancer cells [17]. In the same study, four genes were implicated in the response of cancer cells to tumoricidal agents (e.g. imatinib). Several similar screens on drug sensitivity have led to the identification of cancer-associated genes, e.g., *ACRBP*, *TUBGCP2*, and *MAD2* in breast cancer [62]. Combining RNAi screen and other genomic tools (e.g. SNP array, aCGH, and SAGE data analysis), a recent study identified IKBKE kinase as a breast cancer oncogene [39]. This discovery could lead to the development of pharmaceutical inhibitors that block activity of IKBKE kinase in breast cancer.

Metabolic syndrome, such as obesity, can increase the risk of developing cardiovascular disease and diabetes. However, the underlying molecular mechanisms are far from clear. To understand how adiposity is regulated in *Drosophila*, an *in vivo* genome-wide RNAi screen was reported recently. In this study, transgenic RNAi lines corresponding to 10,489 distinct open reading frames were used in RNAi screen. Tissue-specific gene inactivation for 500 candidates identified from the first screen was further tested. This study reveals hedgehog signaling as one of major adipocyte regulators [18]. To test whether hedgehog signaling plays any role in mammalian adipose tissue, mutant mice were generated to activate hedgehog in adipocytes. Activation of hedgehog in mice adipose tissues resulted in a dramatic loss of white fat compartments (but not brown) by directly blocking differentiation of white adipocytes. These results support the idea that white and brown adipocytes are derived from distinct precursor cells. Interestingly, glucose tolerance and insulin sensitivity remained normal in mutant mice. This suggests that modulating hedgehog signaling can reduce lipid accumulation in white adipose tissue, while maintain a fully functional brown adipose tissue. Since it has been suggested that functional brown adipose tissue represents a potent therapeutic target for obesity control, novel adiposity regulators (e.g. hedgehog signaling) will be developed as obesity drug targets in the near future.

3.3. Agricultural applications

Initially, it's believed that systemic RNAi is a unique feature in worms, until this idea was tested in insect species other than *Drosophila*. The first systemic RNAi in insects was reported in the red flour beetle, *Tribolium castaneum* [63]. Injection of dsRNA for bristle-forming gene, *Tc-achaete-scute* (*Tc-ASH*), resulted in bristle loss phenotype. Following this study, it was discovered that systemic RNAi via dsRNA injection works in many insect species, including mosquitoes [64], honey bees [65], aphids [66], termites [67], etc. Therefore RNAi became a useful tool in functional genomics studies in many non-model insect species, especially those economically important ones.

In 2007, two breakthrough studies described the technology on pest control through feeding transgenic plant expressing dsRNA [68-69]. It is the first evident to show RNAi can be used as a potential pest control strategy for crop protection and feeding RNAi works in certain insect group just like worms. In these studies, initially a list of potential target genes were chosen and dsRNAs against target genes were synthesized *in vitro* and mixed with artificial diet. RNAi for several target genes results in larval growth arrest and lethality. Next, transgenic plant was engineered to produce dsRNA against genes whose inactivation results in strong RNAi response. Such genes include V-type ATPase A in western corn rootworm and cytochrome P450 (CYP6AE14) in cotton bollworm. These results provide strong evidence to support the feasibility of using RNAi in pest control and crop protection. Recently, feeding RNAi was also demonstrated in termites [67]. Feeding on cellulose disks soaked with dsRNA against digestive cellulose enzyme and hexamerin storage protein caused reduction in termite fitness and increased mortality. This study opened a new way for termite control using feed RNAi technology combining with a bait system. Although developing RNAi-based pest control approach is still at early stage and it is not as effective as current crop protection technology (e.g. *Bacillus thuringiensis* (Bt) toxin), RNAi will provide an alternative strategy for the future pest management.

3.4. A case study: Large-scale GPCR RNAi screen for novel pesticide target discovery

The G protein-coupled receptors (GPCRs) belong to the largest superfamily of integral cell membrane proteins and play crucial roles in physiological processes including behavior, development and reproduction. About 1-2% of all genes in an insect genome code for GPCRs. Whole genome sequencing identified about 200 GPCRs in *Drosophila* and 276 GPCRs in African malaria mosquito, *Anopheles gambiae*. Currently, there is not a commercial insecticide that targets GPCR. The red flour beetle, *T. castaneum* is one of the worldwide stored product pests. The genome of *T. castaneum* has been sequenced in 2008 [70], which offers great opportunities for the studies on functional genomics and the identification of targets for pest control. In one recent study [71], 111 non-sensory GPCRs were annotated from the beetle *T. castaneum* genome. To discover potential GPCRs as pesticide targets, a large-scale RNAi screen was performed by injecting dsRNA into developing larvae. The outline of this study is shown in Figure. 1. In this study, eight GPCRs were found involved in larval growth, molting and metamorphosis. The identified GPCRs may serve as potential insecticide targets for controlling *T. castaneum* and other related pest species.

In this GPCR RNAi study [71], 111 annotated *T. castaneum* GPCRs were classified into four different families based on conserved domain prediction program: Class A, Rhodopsin-like receptor; Class B, Secretin receptor-like; Class C, Metabotropic glutamate receptor-like and Class D, Atypical GPCRs. In summary, there are 74 Rhodopsin-like GPCRs, 19 Secretin receptor-like GPCRs, 11 Metabotropic glutamate receptor-like GPCRs, and 7 Atypical GPCRs. Rhodopsin-like GPCR family contains 20 biogenic amine receptors, 42 peptide receptors, four glycoprotein hormone receptors and one purine receptors.

A large-scale GPCR RNAi screen was then conducted by injecting dsRNA for 111 *T. castaneum* GPCRs into one-day-old final instar larvae. Mortality and development defects of dsRNA injected insects were recorded every 2-3 days until adult eclosion. This screen identified 12 GPCRs that effect growth and development. Among 12 GPCRs identified there are biogenic amine receptor (TC007490/D2R), peptide receptors (TC013945/CcapR, TC012493/ETHR, TC004716 and TC006805), and protein hormone receptors (TC008163/bursicon receptor and TC009127/glycoprotein hormone-like receptor). Silencing of genes coding for four GPCRs (TC012521/stan, TC009370/mthl and TC001872/Cirl) in Class B and two GPCRs (TC014055/fz and TC005545/smo) in Class D also caused severe mortality (Table. 3). DsRNA-mediated knockdown for eight GPCRs caused more than 90% mortality after dsRNA injection. Interestingly, RNAi for one of the GPCRs, dopamine-2 like receptor (TC007490), resulted in high lethality during early larval stage. In *Drosophila*, dopamine-2 like receptor (D2R) is one of the genes highly expressed in head and brain (http://www.flyatlas.org/) and D2R RNAi flies with reduced D2R expression show significantly decreased locomotor activity (Draper et al. 2007). Since TC007490/D2R RNAi beetles died during the larval stage, TC007490/D2R might be playing a critical role in the growth and development of beetle larvae by modulating neuronal development and locomotor activity as reported in *D. melanogaster*. Collectively, the RNAi screen in *T. castaneum* has provided useful information and it has also been proven to be a nice model system for future pesticide screen.

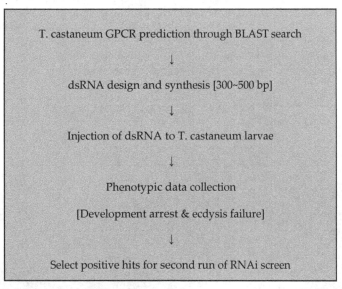

Figure 1. The outline of GPCR RNAi screen in *T. castaneum*

Beside mortality, RNAi for eight GPCRs also resulted in severe developmental arrest and ecdysis failure, including recently characterized bursicon receptor [72]. Interestingly, the majority of insects injected with TC007490/D2R dsRNA was not able to molt to the pupal stage and died during the larval stage. Only, a few larvae injected with TC007490/D2R dsRNA were able to reach quiescent stage (a non-feeding prepupal stage, about 96 hr after ecdysis into final instar), suggesting that this gene may play an important role during larval growth and development rather than molting and metamorphosis. In contrast, most of the insects injected with TC001872/Cirl dsRNA entered the quiescent stage and died during this stage. About 40% of the insect injected TC001872/Cirl dsRNA were able to molt to the pupal stage and eventually died during the early pupal stage. The majority of insects injected with TC012521/stan dsRNA was not able to complete adult eclosion and died during pharate adult stage. Interestingly, TC014055/fz and TC009370/mthl RNAi caused an arrest in both larval-pupal and pupal-adult ecdysis, suggesting that they may play important roles in the regulation of ecdysis behavior. In contrast, insects injected with TC009370/mthl dsRNA were arrested at the late phase of larval-pupal and pupal-adult ecdysis. The majority of insects injected with TC005545/smo dsRNA died during the early pupal stages without showing any ecdysis defects.

The GPCRs identified in this study [71] could be served as potential pesticide targets, which can be used in small molecule screen, or the development of RNAi-based pesticides. Among the identified GPCRs, many of them belong to classic GPCR families, e.g. biogenic amine receptors (TC007490 /D2R and TC011960/5-HTR) and neuropeptide receptors (TC009127/glycoprotein hormone-like receptor). These GPCRs, which are activated by small molecules, can be used as potential tar-gets for novel pesticide development. On the other hand, it may not be possible to apply small molecule ligands for pest management through targeting identified atypical GPCRs (e.g. TC014055 / fz and TC005545 / smo) whose ligands tend to be larger proteins. However, it should be possible to develop a RNAi-based pest control strategy through ingestion of specific dsRNA targeting atypical GPCRs as well as classical GPCRs [73].

Class	Official ID	First Screen		Second Screen	
		Larva Mortality	Pupa Mortality	Larva Mortality	Pupa Mortality
/	*malE* [f]	6.7%	0.0%	2.4%	2.4%
Class A Rhodopsin-like	TC007490	64.3%	35.7%	100.0%	0.0%
	TC008163	10.0%	90.0%	21.2%	75.8%
	TC009127	50.0%	16.7%	40.0%	0.0%
	TC006805	0.0%	62.5%	9.1%	54.5%
	TC013945	0.0%	100.0%	42.1%	52.6%
	TC012493	20.0%	60.0%	*	*
	TC004716	0.0%	41.7%	38.9%	22.2%
Class B Secretin receptor-like	TC001872	55.6%	44.4%	68.4%	31.6%
	TC009370	0.0%	90.0%	42.9%	57.1%
	TC012521	0.0%	90.0%	31.6%	68.4%
Class D Atypical GPCRs	TC014055	0.0%	100.0%	60.0%	40.0%
	TC005545	0.0%	92.3%	46.7%	53.3%

[f]: E. coli malE gene is used as a negative control.

Table 3. Summary of RNAi for 12 GPCRs in *T. castaneum*. Asterisk indicates that RNAi for TC012493 was not carried out at the second screen.

4. Conclusion

Genome-wide RNAi screen is a powerful technique for studying gene functions, deciphering complex phenotypes, and identifying novel drug targets. It opens up a whole new field that allows researchers to explore new modulators in classical signaling pathways, new mechanisms underlying basic biological functions, and new drug targets of human diseases. An increasing number of genome-wide RNAi screens have been successfully conducted for all kinds of novel discoveries. Although the off-target effects and other false discovery issues still remain, RNAi screen technique will be greatly improved as the development of new RNAi libraries and image detection instruments. Most importantly, as our understanding of RNAi pathway continues to grow, we will be able to design more specific and effective RNAi tools for genome-wide RNAi screen. There is no doubt that, through genome-wide RNAi screens, we will gain more insights into complex signaling networks and molecular mechanism of diseases in the near future, which will eventually lead to the discovery of novel therapeutic drug and crop protection reagents.

Author details

Hua Bai
Department of Ecology and Evolutionary Biology, Brown University, USA

Acknowledgement

I'd like to thank Subba R. Palli and Ping Kang for valuable comments on the manuscript, and Ellison Medical Foundation/AFAR postdoctoral fellowship for the financial support.

5. References

[1] Fire A, Xu S, Montgomery MK, Kostas SA, Driver SE, Mello CC. Potent and specific genetic interference by double-stranded RNA in Caenorhabditis elegans. Nature. 1998 Feb 19;391(6669):806-11.

[2] Mello CC, Conte D, Jr. Revealing the world of RNA interference. Nature. 2004 Sep 16;431(7006):338-42.

[3] Zamore PD. RNA interference: big applause for silencing in Stockholm. Cell. 2006 Dec 15;127(6):1083-6.

[4] Ratcliff F, Harrison BD, Baulcombe DC. A similarity between viral defense and gene silencing in plants. Science. 1997 Jun 6;276(5318):1558-60.

[5] Mourrain P, Beclin C, Elmayan T, Feuerbach F, Godon C, Morel JB, et al. Arabidopsis SGS2 and SGS3 genes are required for posttranscriptional gene silencing and natural virus resistance. Cell. 2000 May 26;101(5):533-42.

[6] Voinnet O, Lederer C, Baulcombe DC. A viral movement protein prevents spread of the gene silencing signal in Nicotiana benthamiana. Cell. 2000 Sep 29;103(1):157-67.

[7] Bentwich I, Avniel A, Karov Y, Aharonov R, Gilad S, Barad O, et al. Identification of hundreds of conserved and nonconserved human microRNAs. Nat Genet. 2005 Jul;37(7):766-70.

[8] Reinhart BJ, Slack FJ, Basson M, Pasquinelli AE, Bettinger JC, Rougvie AE, et al. The 21-nucleotide let-7 RNA regulates developmental timing in Caenorhabditis elegans. Nature. 2000 Feb 24;403(6772):901-6.

[9] Arasu P, Wightman B, Ruvkun G. Temporal regulation of lin-14 by the antagonistic action of two other heterochronic genes, lin-4 and lin-28. Genes Dev. 1991 Oct;5(10):1825-33.

[10] Esau C, Davis S, Murray SF, Yu XX, Pandey SK, Pear M, et al. miR-122 regulation of lipid metabolism revealed by in vivo antisense targeting. Cell Metab. 2006 Feb;3(2):87-98.

[11] Trajkovski M, Hausser J, Soutschek J, Bhat B, Akin A, Zavolan M, et al. MicroRNAs 103 and 107 regulate insulin sensitivity. Nature. 2011 Jun 30;474(7353):649-53.

[12] Mencia A, Modamio-Hoybjor S, Redshaw N, Morin M, Mayo-Merino F, Olavarrieta L, et al. Mutations in the seed region of human miR-96 are responsible for nonsyndromic progressive hearing loss. Nat Genet. 2009 May;41(5):609-13.

[13] He L, Thomson JM, Hemann MT, Hernando-Monge E, Mu D, Goodson S, et al. A microRNA polycistron as a potential human oncogene. Nature. 2005 Jun 9;435(7043):828-33.

[14] Consortium CeS. Genome sequence of the nematode C. elegans: a platform for investigating biology. Science. 1998 Dec 11;282(5396):2012-8.

[15] Adams MD, Celniker SE, Holt RA, Evans CA, Gocayne JD, Amanatides PG, et al. The genome sequence of Drosophila melanogaster. Science. 2000 Mar 24;287(5461):2185-95.

[16] Friedman A, Perrimon N. A functional RNAi screen for regulators of receptor tyrosine kinase and ERK signalling. Nature. 2006 Nov 9;444(7116):230-4.

[17] Luo B, Cheung HW, Subramanian A, Sharifnia T, Okamoto M, Yang X, et al. Highly parallel identification of essential genes in cancer cells. Proc Natl Acad Sci U S A. 2008 Dec 23;105(51):20380-5.

[18] Pospisilik JA, Schramek D, Schnidar H, Cronin SJ, Nehme NT, Zhang X, et al. Drosophila genome-wide obesity screen reveals hedgehog as a determinant of brown versus white adipose cell fate. Cell. 2010 Jan 8;140(1):148-60.

[19] Brass AL, Dykxhoorn DM, Benita Y, Yan N, Engelman A, Xavier RJ, et al. Identification of host proteins required for HIV infection through a functional genomic screen. Science. 2008 Feb 15;319(5865):921-6.

[20] Hamilton B, Dong Y, Shindo M, Liu W, Odell I, Ruvkun G, et al. A systematic RNAi screen for longevity genes in C. elegans. Genes Dev. 2005 Jul 1;19(13):1544-55.

[21] Hansen M, Hsu AL, Dillin A, Kenyon C. New genes tied to endocrine, metabolic, and dietary regulation of lifespan from a Caenorhabditis elegans genomic RNAi screen. PLoS Genet. 2005 Jul;1(1):119-28.

[22] Mohr S, Bakal C, Perrimon N. Genomic screening with RNAi: results and challenges. Annu Rev Biochem. 2010;79:37-64.

[23] Perrimon N, Ni JQ, Perkins L. In vivo RNAi: today and tomorrow. Cold Spring Harb Perspect Biol. 2010 Aug;2(8):a003640.

[24] Ramadan N, Flockhart I, Booker M, Perrimon N, Mathey-Prevot B. Design and implementation of high-throughput RNAi screens in cultured Drosophila cells. Nat Protoc. 2007;2(9):2245-64.

[25] Clemens JC, Worby CA, Simonson-Leff N, Muda M, Maehama T, Hemmings BA, et al. Use of double-stranded RNA interference in Drosophila cell lines to dissect signal transduction pathways. Proc Natl Acad Sci U S A. 2000 Jun 6;97(12):6499-503.

[26] Hammond SM, Bernstein E, Beach D, Hannon GJ. An RNA-directed nuclease mediates post-transcriptional gene silencing in Drosophila cells. Nature. 2000 Mar 16;404(6775):293-6.

[27] Kamentsky L, Jones TR, Fraser A, Bray MA, Logan DJ, Madden KL, et al. Improved structure, function and compatibility for CellProfiler: modular high-throughput image analysis software. Bioinformatics. 2011 Apr 15;27(8):1179-80.

[28] Kiger AA, Baum B, Jones S, Jones MR, Coulson A, Echeverri C, et al. A functional genomic analysis of cell morphology using RNA interference. J Biol. 2003;2(4):27.

[29] Bjorklund M, Taipale M, Varjosalo M, Saharinen J, Lahdenpera J, Taipale J. Identification of pathways regulating cell size and cell-cycle progression by RNAi. Nature. 2006 Feb 23;439(7079):1009-13.

[30] Sims D, Duchek P, Baum B. PDGF/VEGF signaling controls cell size in Drosophila. Genome Biol. 2009;10(2):R20.

[31] Gonsalves FC, Klein K, Carson BB, Katz S, Ekas LA, Evans S, et al. An RNAi-based chemical genetic screen identifies three small-molecule inhibitors of the Wnt/wingless signaling pathway. Proc Natl Acad Sci U S A. 2011 Apr 12;108(15):5954-63.

[32] Seyhan AA, Varadarajan U, Choe S, Liu Y, McGraw J, Woods M, et al. A genome-wide RNAi screen identifies novel targets of neratinib sensitivity leading to neratinib and paclitaxel combination drug treatments. Mol Biosyst. 2011 Jun;7(6):1974-89.

[33] Zhu YX, Tiedemann R, Shi CX, Yin H, Schmidt JE, Bruins LA, et al. RNAi screen of the druggable genome identifies modulators of proteasome inhibitor sensitivity in myeloma including CDK5. Blood. 2011 Apr 7;117(14):3847-57.

[34] Kulkarni MM, Booker M, Silver SJ, Friedman A, Hong P, Perrimon N, et al. Evidence of off-target effects associated with long dsRNAs in Drosophila melanogaster cell-based assays. Nat Methods. 2006 Oct;3(10):833-8.

[35] Booker M, Samsonova AA, Kwon Y, Flockhart I, Mohr SE, Perrimon N. False negative rates in Drosophila cell-based RNAi screens: a case study. BMC Genomics. 2011;12:50.

[36] Bakal C, Linding R, Llense F, Heffern E, Martin-Blanco E, Pawson T, et al. Phosphorylation networks regulating JNK activity in diverse genetic backgrounds. Science. 2008 Oct 17;322(5900):453-6.

[37] Tu Z, Argmann C, Wong KK, Mitnaul LJ, Edwards S, Sach IC, et al. Integrating siRNA and protein-protein interaction data to identify an expanded insulin signaling network. Genome Res. 2009 Jun;19(6):1057-67.

[38] Berns K, Hijmans EM, Mullenders J, Brummelkamp TR, Velds A, Heimerikx M, et al. A large-scale RNAi screen in human cells identifies new components of the p53 pathway. Nature. 2004 Mar 25;428(6981):431-7.

[39] Boehm JS, Zhao JJ, Yao J, Kim SY, Firestein R, Dunn IF, et al. Integrative genomic approaches identify IKBKE as a breast cancer oncogene. Cell. 2007 Jun 15;129(6):1065-79.

[40] Sessions OM, Barrows NJ, Souza-Neto JA, Robinson TJ, Hershey CL, Rodgers MA, et al. Discovery of insect and human dengue virus host factors. Nature. 2009 Apr 23;458(7241):1047-50.

[41] Konig R, Stertz S, Zhou Y, Inoue A, Hoffmann HH, Bhattacharyya S, et al. Human host factors required for influenza virus replication. Nature. 2010 Feb 11;463(7282):813-7.

[42] Meacham CE, Ho EE, Dubrovsky E, Gertler FB, Hemann MT. In vivo RNAi screening identifies regulators of actin dynamics as key determinants of lymphoma progression. Nat Genet. 2009 Oct;41(10):1133-7.

[43] Bric A, Miething C, Bialucha CU, Scuoppo C, Zender L, Krasnitz A, et al. Functional identification of tumor-suppressor genes through an in vivo RNA interference screen in a mouse lymphoma model. Cancer Cell. 2009 Oct 6;16(4):324-35.

[44] Fraser AG, Kamath RS, Zipperlen P, Martinez-Campos M, Sohrmann M, Ahringer J. Functional genomic analysis of C. elegans chromosome I by systematic RNA interference. Nature. 2000 Nov 16;408(6810):325-30.

[45] Dillin A, Hsu AL, Arantes-Oliveira N, Lehrer-Graiwer J, Hsin H, Fraser AG, et al. Rates of behavior and aging specified by mitochondrial function during development. Science. 2002 Dec 20;298(5602):2398-401.

[46] Lee SS, Lee RY, Fraser AG, Kamath RS, Ahringer J, Ruvkun G. A systematic RNAi screen identifies a critical role for mitochondria in C. elegans longevity. Nat Genet. 2003 Jan;33(1):40-8.

[47] Wang MC, O'Rourke EJ, Ruvkun G. Fat metabolism links germline stem cells and longevity in C. elegans. Science. 2008 Nov 7;322(5903):957-60.

[48] Parry DH, Xu J, Ruvkun G. A whole-genome RNAi Screen for C. elegans miRNA pathway genes. Curr Biol. 2007 Dec 4;17(23):2013-22.

[49] Maeda I, Kohara Y, Yamamoto M, Sugimoto A. Large-scale analysis of gene function in Caenorhabditis elegans by high-throughput RNAi. Curr Biol. 2001 Feb 6;11(3):171-6.

[50] Ni JQ, Markstein M, Binari R, Pfeiffer B, Liu LP, Villalta C, et al. Vector and parameters for targeted transgenic RNA interference in Drosophila melanogaster. Nat Methods. 2008 Jan;5(1):49-51.

[51] Saj A, Arziman Z, Stempfle D, van Belle W, Sauder U, Horn T, et al. A combined ex vivo and in vivo RNAi screen for notch regulators in Drosophila reveals an extensive notch interaction network. Dev Cell. 2010 May 18;18(5):862-76.

[52] Cronin SJ, Nehme NT, Limmer S, Liegeois S, Pospisilik JA, Schramek D, et al. Genome-wide RNAi screen identifies genes involved in intestinal pathogenic bacterial infection. Science. 2009 Jul 17;325(5938):340-3.

[53] Neely GG, Kuba K, Cammarato A, Isobe K, Amann S, Zhang L, et al. A global in vivo Drosophila RNAi screen identifies NOT3 as a conserved regulator of heart function. Cell. 2010 Apr 2;141(1):142-53.

[54] Neely GG, Hess A, Costigan M, Keene AC, Goulas S, Langeslag M, et al. A genome-wide Drosophila screen for heat nociception identifies alpha2delta3 as an evolutionarily conserved pain gene. Cell. 2010 Nov 12;143(4):628-38.

[55] Neumuller RA, Richter C, Fischer A, Novatchkova M, Neumuller KG, Knoblich JA. Genome-wide analysis of self-renewal in Drosophila neural stem cells by transgenic RNAi. Cell Stem Cell. 2011 May 6;8(5):580-93.

[56] Fernandes C, Rao Y. Genome-wide screen for modifiers of Parkinson's disease genes in Drosophila. Mol Brain. 2011;4:17.

[57] Lee SS. Whole genome RNAi screens for increased longevity: important new insights but not the whole story. Exp Gerontol. 2006 Oct;41(10):968-73.

[58] Kenyon C, Chang J, Gensch E, Rudner A, Tabtiang R. A C. elegans mutant that lives twice as long as wild type. Nature. 1993 Dec 2;366(6454):461-4.

[59] Kaeberlein M, Powers RW, 3rd, Steffen KK, Westman EA, Hu D, Dang N, et al. Regulation of yeast replicative life span by TOR and Sch9 in response to nutrients. Science. 2005 Nov 18;310(5751):1193-6.

[60] Apfeld J, O'Connor G, McDonagh T, DiStefano PS, Curtis R. The AMP-activated protein kinase AAK-2 links energy levels and insulin-like signals to lifespan in C. elegans. Genes Dev. 2004 Dec 15;18(24):3004-9.

[61] Paik D, Jang YG, Lee YE, Lee YN, Yamamoto R, Gee HY, et al. Misexpression screen delineates novel genes controlling Drosophila lifespan. Mech Ageing Dev. 2012 Feb 24.

[62] Whitehurst AW, Bodemann BO, Cardenas J, Ferguson D, Girard L, Peyton M, et al. Synthetic lethal screen identification of chemosensitizer loci in cancer cells. Nature. 2007 Apr 12;446(7137):815-9.

[63] Tomoyasu Y, Denell RE. Larval RNAi in Tribolium (Coleoptera) for analyzing adult development. Dev Genes Evol. 2004 Nov;214(11):575-8.

[64] Sanchez-Vargas I, Travanty EA, Keene KM, Franz AW, Beaty BJ, Blair CD, et al. RNA interference, arthropod-borne viruses, and mosquitoes. Virus Res. 2004 Jun 1;102(1):65-74.

[65] Nunes FM, Simoes ZL. A non-invasive method for silencing gene transcription in honeybees maintained under natural conditions. Insect Biochem Mol Biol. 2009 Feb;39(2):157-60.

[66] Shakesby AJ, Wallace IS, Isaacs HV, Pritchard J, Roberts DM, Douglas AE. A water-specific aquaporin involved in aphid osmoregulation. Insect Biochem Mol Biol. 2009 Jan;39(1):1-10.

[67] Zhou X, Wheeler MM, Oi FM, Scharf ME. RNA interference in the termite Reticulitermes flavipes through ingestion of double-stranded RNA. Insect Biochem Mol Biol. 2008 Aug;38(8):805-15.

[68] Baum JA, Bogaert T, Clinton W, Heck GR, Feldmann P, Ilagan O, et al. Control of coleopteran insect pests through RNA interference. Nat Biotechnol. 2007 Nov;25(11):1322-6.

[69] Mao YB, Cai WJ, Wang JW, Hong GJ, Tao XY, Wang LJ, et al. Silencing a cotton bollworm P450 monooxygenase gene by plant-mediated RNAi impairs larval tolerance of gossypol. Nat Biotechnol. 2007 Nov;25(11):1307-13.

[70] Richards S, Gibbs RA, Weinstock GM, Brown SJ, Denell R, Beeman RW, et al. The genome of the model beetle and pest Tribolium castaneum. Nature. 2008 Apr 24;452(7190):949-55.

[71] Bai H, Zhu F, Shah K, Palli SR. Large-scale RNAi screen of G protein-coupled receptors involved in larval growth, molting and metamorphosis in the red flour beetle. BMC Genomics. 2011;12:388.

[72] Bai H, Palli SR. Functional characterization of bursicon receptor and genome-wide analysis for identification of genes affected by bursicon receptor RNAi. Dev Biol. 2010 Aug 1;344(1):248-58.

[73] Price DR, Gatehouse JA. RNAi-mediated crop protection against insects. Trends Biotechnol. 2008 Jul;26(7):393-400.

Chapter 6

How RNA Interference Combat Viruses in Plants

Bushra Tabassum, Idrees Ahmad Nasir, Usman Aslam and Tayyab Husnain

Additional information is available at the end of the chapter

1. Introduction

RNA mediated silencing technology has now become the tool of choice for induction of virus resistance in plants. A significant feature of this technology is the presence of double-stranded RNA (dsRNA), which is not only the product of RNA silencing but also the potent triggers of RNA interference (RNAi). Upon RNAi induction, these dsRNAs are diced into short RNA fragments termed as small interfering RNAs (siRNAs), which are hallmarks of RNAi. Considerable resistance in transgenic plants against viruses can be created by exploiting the phenomenon of RNAi. In the current chapter, generation of potato virus Y (*PVY*) resistant potato and sugarcane mosaic virus (*SCMV*) resistant sugarcane by CEMB has been quoted as an example.

We are in the dawn of a new age in functional genomics driven by RNAi methods. RNA interference (RNAi) refers to a post-transcriptional process triggered by the introduction of double-stranded RNA (dsRNA) which leads to gene silencing in a sequence-specific manner. It is one of the most exciting discoveries of the past decade in functional genomics and is rapidly becoming an important method for analyzing gene functions in eukaryotes and holds promise for the development of therapeutic gene silencing and which is therefore currently the most widely used gene-silencing technique in functional genomics.

2. Need for resistance

Agriculture sector of any country strengthen the economy by contributing in its gross domestic product (GDP). In Pakistan, the major agricultural crops include cotton, wheat, rice, sugarcane, potato and tomato etc. All of the above mentioned crops has great potential for yield and contribute 24 % in gross domestic product (GDP) of Pakistan economy [1]. There is a major gap in the actual yield potential of each crop with respect to its harvested yield, possible reasons include disease attack, environmental damages and in some cases lack of quality seed. Disease attack as being the most common cause include infections by pathogens like viruses, bacteria and insects etc. Viruses can cause most devastating effects due to their systemic infections and hence decrease crop productivity in primary infection and reduce seed quality for subsequent use through persistent infection. Viral epidemics are often associated with the emergence of a new form of the viral strain or a new form of vector.

Viruses are a major threat to agriculture all over the world. Up till now, more than 1200 plant viruses have been reported which include 250 of those viruses that cause significant losses in crop yield [2]. In nature, viral particles exist as obligate parasites which consist of hereditary material packed in a thick layered coat and completely depend on host cell throughout their life cycle. Viruses utilizes host resources like nucleic acid, amino acids and certain proteins for their replication and survival, thus disturbing host plant metabolism to a considerable extent. Most of the infecting plant viruses are ssRNA viruses like sugarcane mosaic virus, potato virus Y etc. In an infected plant, virus accumulation goes higher with increased progeny rate through its replication. The spread of the virus in a plant is achieved through its movement from infected cell to healthy one via plasmodesmata while long distance movement occurs through phloem. Entry of virus in plant usually occurs through physical injury like wound etc or via certain viral vectors like aphid, fly etc.

Cotton (Gossypium hirsutum) as being commonly known as 'white gold' is an important cash crop in many developing countries including Pakistan. It is a natural fibre and has many uses in industries. It accounts for 8.2 percent of the value added in agriculture and about 2 percent to GDP of Pakistan. The yield of the crop is severely affected by the viruses including geminiviruses (leaf crumple and leaf curl) and tobacco streak virus etc. These viruses can cause severe losses when infections occur on young plants; some infect cotton yield while others affect lint quality as well [3].

Potato (Solanum tuberosum) is the world's major food crop and is one of the leading vegetables. Viruses are a serious problem, not only because of effects caused by primary infection, but also because the crop is vegetatively propagated and the viruses are transmitted through the tubers to subsequent generations. Potato virus Y (PVY) is probably the most damaging and widespread virus of potato and is found wherever potato crops are grown, where losses are reported upto 10 to 90% [4]. PVY is transmitted through aphids.

Sugarcane (Saccharum spp. hybrid) is among the top 10 food crops of the world, and yearly provides 60% to 70% of the sugar produced around the world [5]. Yield harvested by the farmers of Pakistan is very low whose main cause is mosaic disease of sugarcane which continues to be a potential threat to the sugarcane production. It is a very common disease in all the major sugarcane growing regions, because of the perpetuation of the disease virus through vegetative propagules. Sugarcane mosaic virus (SCMV) is reported to infect sugarcane naturally and can cause severe losses to the farmers and lesser production to the industry [6,7]. Aphids are the vector for transmission of the disease. Seed produced by infected cane can also transmit the disease.

Tomato (*Lycopersicum esculentum*) ranks among the most widely grown vegetables all over the world. In general, viral diseases are not a routine problem in most tomato plantings but incidence of some viruses including tomato spotted wilt virus, tomato leaf rolling, tobacco mosaic virus and single and double virus streak viruses has devastating impact on crop yield with losses of upto 100% have also been reported [8]. Tobacco mosaic virus is one of the most stable viruses known because it is able to survive in dried plant debris as long as 100 years.

Various control measures have been taken to overcome losses caused by plant viruses which are expensive and also inadequate. Biotechnologists have developed and adopted several strategies for virus resistance in crop plants. These include cross protection, pathogen derived resistance and more recently RNA interference. Therefore, the development of virus resistant varieties seems the only economically feasible way to control viruses [9]. Today, use of resistant varieties has been advocated as the most promising and least expensive method of viral disease suppression. With the appealing results of RNAi in silencing target genes of attacking virus, RNAi seems to have potential for creating virus resistant crops.

A plant is said to be resistant if it has the ability to suppress viral disease symptoms by inhibiting its replication or by blocking the virus expression. Resistance mechanism in plants may be either protein mediated or RNA mediated, however final outcome of both is reduced accumulation of virus in the host plant. Acquired resistance could be either high almost reaching immunity with no disease symptoms or moderate to low where mild symptoms of particular viral disease can be seen. In contrast, when a viral infected plant shows normal growth rate with a good yield along with milder symptoms of disease, it is said to be a tolerant plants. In this case, the host plant supports multiplication of virus rather than blocking its replication [10,11].

3. History of virus resistance in transgenic plants

When a plant encounters virus, it reacts naturally through hypersensitive response (HR) and extreme resistance response (ER) which induces the production of secondary metabolites termed as response elements in plants. These response elements include elevated levels of ethylene, jasmonic acid, salicylic acid, nitric oxide and increased rate of ion flux, in combination these factors block the virus entry and /or helps eliminate the virus (figure 1).

The acquired virus resistance mechanisms in plants are of two types: a) gene silencing independent virus resistance and b) gene silencing dependable virus resistance via Post Transcriptional Gene Silencing (PTGS). The first includes coat protein-mediated, movement protein-mediated and replicase protein-mediated resistance, while second includes pathogen-derived resistance, antisense RNA mediated resistance and RNA-mediated resistance. PTGS is an evolutionary conserved mechanism in plants against potential harms by viruses and transposons. In this process, a plant defends itself by exploiting the requirement of plant RNA viruses to replicate using a double-stranded, replicative intermediate (dsRNA). The double-stranded RNA produced is cleaved into approximately

21 nucleotide fragments by the Dicer enzyme [12]. Evidence suggest that transgene loci and RNA viruses can generate double-stranded RNAs which are similar in sequence to the transcribed region of target genes, which further undergo endonucleolytic cleavage to generate small interfering RNAs (siRNA) that promote degradation of cognate RNAs.

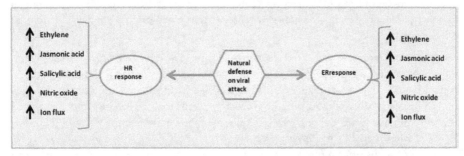

Figure 1. Natural response of plants against viral attack, where production of secondary metabolites cause extreme or moderate resistance.

The first approach made by plant agronomists was the inoculation of susceptible plant with a milder strain of the target virus. This technique was named as **cross protection** and was employed on crops like tomato, papaya and citrus [13-15]. Scientists were met with success as considerable resistance was achieved in transgenic plants through employment of this approach but the success was accompanied with a major drawback that the milder strain of the virus providing protection to one crop may cause serious diseases on varieties growing nearby.

To compensate the drawback of cross protection, **pathogen derived resistance** (PDR) based strategies were employed. These are based on the insertion of resistant genes that are derived from the pathogen (virus) into the host plant. Resistance was achieved by expressing viral genes in plants including coat protein, movement protein and replicase protein gene, each of them targets at a step crucial to virus replication. Coat protein gene is responsible for viral uncoating and is involved in virus replication [16], movement protein is crucial for cell to cell movement of the infecting virus [17] whereas the Rep protein is involved in virus replication and its genome integrity [18,19]. Resistance was either due to protein accumulation (coat protein mediated resistance, movement protein mediated resistance and replicase protein mediated resistance) or because of accumulation of small RNA sequences (replicase mediated resistance).

Uptill now, scientists have made considerable successful attempts to generate virus resistant transgenic plants by employing PDR concept [20, 21]. For example, virus-resistant potato varieties having *PVY* coat protein (CP) or P1 gene sequences has been reported in numerous studies [22-27]. Biotechnologists employed various genes of *PVY* and have met with mixed success in engineering *PVY* resistant transgenic potato plants [22,28-30,24,31,25,32]. In another study, the presence of the movement protein (pr17 protein) was reported to create resistance in transgenic plants against luteovirus Potato leaf roll virus [33].

Although pathogen derived resistance strategies hold promise for upto 90% resistance against target virus and are being employed still to date but some remarkable and potential threats are also associated with the use of this technology. The major one includes; the expression of a gene fragment derived from virus in transgenic plant confers resistance to particular virus but at the same time also raises environmental safety concerns regarding the constitutive expression of viral genes. It is supposed that infecting virus can interact with expression product in transgenic plants and can potentially modify the biological properties of the existing virus, ultimately leading to creation of new virus species which have novel pathogenic properties, host range and altered transmission specificity. In the initial experiments, the virus resistance was based on protein expression but resistance was neither so stable nor effective as compared to the resistance achieved through RNAi.

Among pathogen derived resistance strategies, **antisense RNA** complementary to part of the viral genome proves to have potential utility for protecting plants from systemic virus infection [34]. Antisense RNAs refer to small untranslatable RNA molecules that pair with a target RNA sequence on homology basis and thereby exert a negative control on interaction of target RNA with other nucleic acids or protein factors. Further, RNase H cause an increase in rate of degradation of double stranded RNA [35]. Antisense RNA technology was quickly adopted by plant researchers because other approaches like homologous recombination and gene-tagging mutagenesis used were based on reverse genetics and also these were not applicable in plants nor these were well developed. This background makes antisense RNA-mediated suppression more powerful tool for transgenic research and also for the development of commercial products [36].

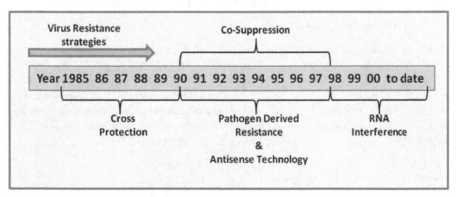

Figure 2. Major milestones in virus resistance strategies drawn to scale, starting form cross protection to RNA-mediated gene suppression.

4. RNA silencing

The development of the concept of pathogen derived resistance gave rise to strategies ranging from coat protein based interference of virus propagation to RNA mediated virus gene silencing. Virus resistance is achieved usually through the antiviral pathways of RNA silencing, a natural defense mechanism of plants against viruses. The experimental approach consists of isolating a segment of the viral genome itself and transferring it into the genome of a susceptible plant. Integrating a viral gene fragment into a host genome does not cause disease (the entire viral genome is needed to cause disease). Instead, the plant's natural antiviral mechanism that acts against a virus by degrading its genetic material in a nucleotide sequence specific manner via a cascade of events involving numerous proteins, including ribonucleases (enzymes that cleave RNA) is activated. This targeted degradation of the genome of an invader virus protects plants from virus infection.

a.   Transcriptional Gene Silencing (TGS)

In plants, silencing is of two types: transcriptional and post transcriptional gene silencing. In both types, the inactivated genes are in trans position as homologous genes upon interaction reside on opposite chromosomes. TGS and PTGS differ from each other with respect to the underlying mechanism they exhibit. TGS requires sequence homology between promoters as compared to PTGS which require homology between coding region of the interacting genes. In TGS, an inactive allele residing on one chromosome can render another allele silenced. The mechanism behind transcriptional gene silencing is suggested to be DNA-DNA interaction which is thought to play an important role [37,38]. In other studies, it was proposed that RNA molecules interact with DNA and subsequently induce DNA methylation which then leads to gene silening [39-42], however it is not clear whether methylation of DNA alone is sufficient for silencing or not. It is proposed that DNA methylation in promoter region has a strong negative effect on interaction of certain transcription factors with promoter. Possible mechanism of TGS is depicted in figure 3.

b.   Post Transcriptional Gene Silencing (PTGS)

'RNA interference' is a conserved mechanism of post transcriptional gene silencing (PTGS). It has rapidly gained favor as a "reverse genetics" tool to knock down the expression of targeted genes in plants. The term RNAi was coined in 1998 by Fire and Mello to describe a gene-silencing phenomenon based on double-stranded RNA [43]. PTGS mechanism controls processes including development, the maintenance of genome stability and defense against molecular parasites (transposons and viruses). Several reports pointed out that PTGS in plants is strictly linked to RNA virus resistance mechanism [44-46].

5. Mechanism of RNAi/PTGS

RNAi (RNA interference) is a natural defense pathway evolved in plants against viruses and potential transposons. It is a cellular pathway in which target sequences are degraded on homology basis at mRNA level by small RNAs, thereby preventing the translation of target RNAs. In plants, two functionally different RNAs; microRNA (miRNA) and small interfering RNA (siRNA), have been characterized. The miRNAs are small 21-26nt long dsRNAs that are genome coded and are endogenous to every cell. Structurally, they comprised of a stem region which is double stranded and a loop region which is single stranded. The miRNAs generated from endogenous hpRNA precursors and are basically involved in the regulation of development [47]. On the other hand, siRNAs are generated from long dsRNA and are involved in defense through RNA interference [48, 49].

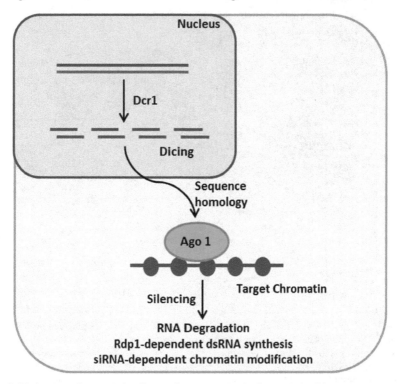

Figure 3. Mechanism of transcriptional gene silencing, active in chromatin modification.

RNAi is an immune system in plants which is directed against viruses [50]. Upon viral attack, long dsRNAs are produced from the replication intermediates of viral RNAs that act as substrate for an endonuclease termed Dicer which is located in the cytosol [51]. Dicer recognizes these dsRNAs and cleave them into duplex siRNA (21-25 nt) [52]. The siRNA duplex comprised of two strands; strand complementary to target mRNA is guide strand and other is passenger strand. The guide strand of short siRNA duplex is incorporated into the RNA-induced silencing complex (RISC) and then siRNA programmed RISC degrade viral RNA. As the RISC complex encounters a foreign mRNA which could be of virus origin, it has two consequences. 1) If the homology of guide strand and target mRNA is 100%, then perfect complement form between them resulting in mRNA cleavage and subsequent degradation or 2) in case of imperfect complement, where few mismatches exist between guide strand of RISC and target mRNA, translation of target mRNA is inhibited (figure 4).

Same mechanism operates in microRNA triggered gene silencing. miRNAs processed from stem loop precursors (shRNA and/or hpRNA) and requires Dicer activity [53] followed by

RISC assembly and subsequent degradation of homologous RNA in a sequence specific manner.

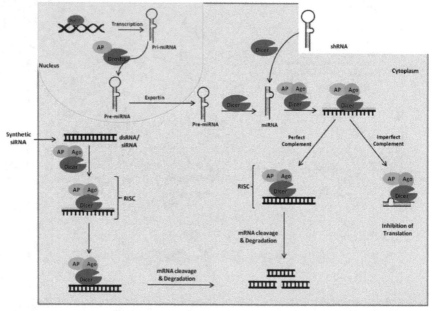

Figure 4. Model for RNA silencing, an ordered biochemical pathway which is triggered by dsRNA of viral origin. The source of dsRNA is either the synthetic siRNA or pre-microRNA. Genome encoded pri-miRNAs are processed by Drosha (an RnaseIII enzyme) into pre-miRNAs which are exported in the cytosol. dsRNA (siRNA or miRNA) subsequently joins Dicer, Ago and some other accessory proteins located in the cytosol forming RISC (RNA induced Silencing Complex). The degree of complementarity between the RNA silencing molecule and its cognate target determines the fate of the mRNA: blocked translation or mRNA cleavage/ degradation.

RISC is a combination of Dicer (an endonuclease enzyme), some accessory proteins namely argonaute (ago1, 4, 6, 9; catalytic endonucleases) and RNA binding proteins (RBP), and some trans-acting RNA-binding proteins (TRBP) [54,55].

Stability of RNAi induced silencing is based on enzymatic methylation of siRNA. This reaction is catalyzed by the enzyme methyltransferase (HEN1) which methylates the siRNA at 3′ end, hereby preventing it from oligouridylation and subsequent degradation [56].

6. Systemic spread of RNAi

When RNAi is induced at one site in an organism including plant, a mobile signal is generated which spread cell to cell and systemically throughout the organism [57-59,43,60,61] and make RNAi response obvious in distant tissues of the plant. This silencing signal moves inside plant either through the intercellular spaces called plasmodesmata or through the phloem as shown in figure 5 [59,60]. Presence of a mobile signal has been proposed to be an integral part in systemic spread of silencing. The first evidence of the presence of a mobile silencing signal came from the study of Agro infiltration assay or particle bombardment in development of transgenic tobacco plants [59,60,62]. Subsequently, in silenced tissues of Agro-infiltrated plants, T-DNA or Agrobacterium was detected which suggests that mobile signal is responsible for propagation of silencing from one tissue to another [59] and this signal can also cross graft junction [59,60,62]. Candidates proposed to be responsible for mobile silencing signal involve siRNAs, Aberrant RNAs and dsRNA [63].

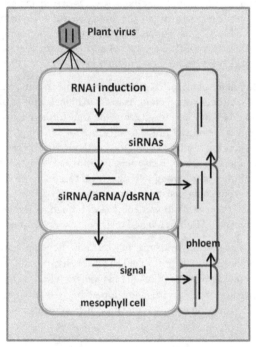

Figure 5. Mobile silencing signal passes from infected cell to healthy cell upon RNAi induction. Candidate of RNA silencing could siRNA, aRNA or dsRNA and travel through plasmodesmata and/or phloem.

Conclusively, transgenic approach mediated by RNAi pre-programmed an existing antiviral defense in plants [21,64-66]. Plant viruses are the strong inducers of RNAi as well as a target. The simplicity and specificity of RNAi has made RNAi a routine tool for the generation of virus resistant crops.

7. Effective RNAi inducers

In general, gene silencing has proven fruitful with both sense- and antisense transgenes in plant cells [67,68]. An RNA molecule that contains a fragment of a sense strand, an antisense strand and a short loop sequence between the fragment making a tight hairpin turn is termed as short hairpin RNA (shRNA) which has the ability to suppress the expression of desired genes via RNA interference [69]. Silencing can be more efficiently achieved by utilizing shRNA cassettes [70-72] which usually include a specific plant promoter and terminator sequences to control the expression of inversely repeated sequences of the dsRNA. Upon subsequent delivery of shRNA cassette in the plant cells, dsRNA molecules comprised of a loop (single-stranded) and a stem region (double-stranded) are formed. Further, stem region is used by Dicer as a substrate and trigger RNAi mechanism [72-74]. RNA silencing mediated by the use of shRNA cassette enforces stable and heritable gene silencing [67] as it utilizes the specific promoter to ensure that the shRNA is always expressed. Another reason which justify that the silencing efficiency can be more powerful when using shRNA cassette is due to the fact that dsRNA are being fed into a later step in the silencing pathway where they act as a substrate for Dicer (RNaseIII like enzyme) and therefore bypasses the step in which dsRNAs need plant encoded RdRps for their production [75].

Practically in development of virus resistant transgenic plants, specific hairpinRNA expression constructs have been designed for transformation. In this strategy, small dsRNAs which are hallmark of PTGS, are produced from the transformed construct and ultimately induce silencing. Scientists have used hpRNA construct for silencing of viral gene in potato and obtained efficient silencing results accompanied with production of siRNA [76]. Similarly, some others have compared various constructs in terms of their silencing potential and confirmed that most efficient and strong silencing in tobacco can be achieved through the expression of an intron containing construct, which trigger PTGS [77]. However, in another study where *PVY* resistant potato plants were obtained through CP gene expression, evidence for existence of both protein and RNA mediated mechanisms was verified [27].

While considering the appealing outcome of RNAi in development of virus resistant transgenic plants as reviewed in this article and the use of hairpin RNA for strong silencing, production of transgenic potato resistant against potato virus Y and sugarcane plants resistant against sugarcane mosaic virus developed by [69] at Centre of Excellence in Molecular Biology (CEMB), University of the Punjab has been quoted as an example in following chapter.

8. Development of *PVY* and *SCMV* resistant transgenic plants

Tabassum *et al.* [79] have developed *PVY* and *SCMV* resistant potato and sugarcane plants respectively through siRNA technology by targeting capsid protein gene of respective virus. In the study, the respective plant was equipped with shRNA cassette that reacts continuously against invading virus specifically, thus resulting in degradation of viral mRNA in a sequence-specific manner. Specialty of this shRNA cassette is that it contained screened siRNA (the one most efficient in *in-vitro* experiments) out of bulk. The 22nt long siRNA was used as core sequence in shRNA cassette while loop sequence and flanking sequences were taken from highly active regulatory microRNA of respective host plant.

Initially, for screening of siRNA out of bulk, a strategy based on transient transfection assay was optimized in mammalian cell line (CHO). mRNA knockdown efficiency of capsid protein gene of target virus was analyzed by real-time PCR. In case of *PVY* capsid gene, one specific siRNA out of a total six was found to be the most effective for knockdown of respective mRNA in transfected CHO cells by up to 80-90%. Data obtained showed that all six siRNAs used reduced the mRNA expression of target gene to some extent but only siRNA1 significantly reduced CP-*PVY* mRNA expression by up to 12.25 fold and, as is clearly shown in figure 6, expression was almost diminished or very faint in cells transfected with siRNA1 as compared to the control where scrambled siRNA was transfected. The remaining siRNA knockdown values were: siRNA 2 - 7x decrease ; siRNA 3 - 8x decrease; siRNA 4 - 10.8 x decrease; siRNA 5 - 9x decrease and siRNA 6 - 10x decrease. These values were based on Ct values obtained from real-time PCR studies [78].

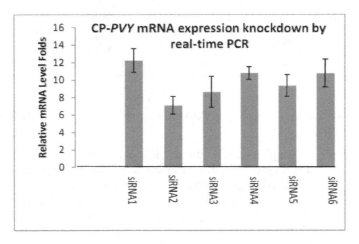

Figure 6. Relative measure of the knockdown of mRNA expression of CP-*PVY* gene in transient transfection assays. Knockdown values are based on relative Ct values obtained in realtime PCR assay; GAPDH was used as internal control to normalize the results.

Similar findings were met when knockdown in mRNA expression of CP-*SCMV* was studied *in-vitro* through transient transfection assays. As clear from figure 7, siRNA1 reduced the mRNA expression of target gene by upto 96%, while inhibition by siRNA2 was 46%, siRNA3 and siRNA4 inhibited target gene mRNA expression upto 50% and 77% respectively (figure 7).

Subsequently, the screened siRNA for both viruses was used in shRNA cassette which is thought to synthesize target specific siRNAs that continuously guard the plant against respective viral attack. shRNA cassette cloned in pCAMBIA1301 vector and transformed in potato and sugarcane through Agrobacterium- and particle bombardment method respectively. Results were compared with control non-transgenic plants. Figure 8 and 9 depicts the results, clearly indicating that in transgenic potato having shRNA1 cassette integrated in them, mRNA knockdown was upto 96% whereas in transgenic potato plants having shRNA4 cassette in them, *PVY* knockdown was upto 57% as compared to control where *PVY* infection was maximum.

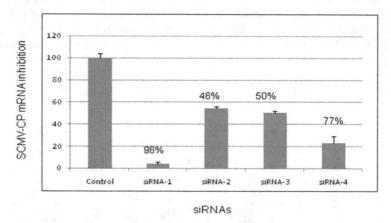

siRNAs

Figure 7. Relative measure of the knockdown of mRNA expression of CP-*SCMV* gene in transient transfection assays.

Similarly, in transgenic sugarcane plants, shRNA1 reduced the mRNA expression of *SCMV* to lesser extent with 30% reduction only while shRNA4 caused maximum knockdown of 95% as compared to the control non-transgenic sugarcane plant.

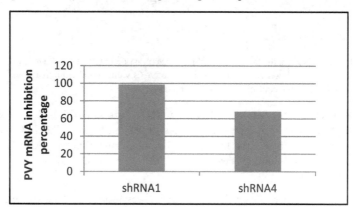

Figure 8. Percentage inhibition in mRNA expression of *PVY* rendered by integrated shRNA1 and shRNA4 cassette in potato plants. Transgenic plants were subjected to bioassay by *PVY* inoculation and RT-PCR was performed 30 days post *PVY* inoculation.

In conclusion, we have developed transgenic potato and sugarcane plants that were highly resistant against *PVY* and *SCMV* infection respectively. This resistance was because of the shRNA cassette integrated in them that is targeted against capsid protein gene of each virus.

These shRNAs are supposed to create long-term targeted gene inhibition in cells and whole plant. Our shRNA construct designing was based on the hypothesis that if we express potentially effective screened siRNA in hairpin form which is further combined with the power of most active regulatory microRNA in respective plant, the level of resistance will be far more effective. Applying this theme, we were able to obtain transgenic potato and sugarcane plants where resistance level against targeted virus was upto immunity.

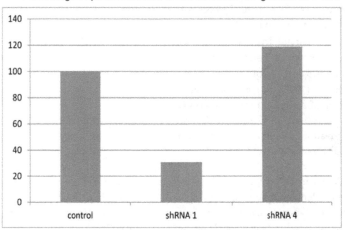

Percentage expression knock down of SCMV CP gene in SPF-234

Figure 9. Percentage inhibition in mRNA expression of *SCMV* rendered by integrated shRNA1 and shRNA4 cassette in sugarcane plants. Transgenic plants were subjected to bioassay by *SCMV* infection and RT-PCR was performed 30 days post inoculation.

One important aspect of this strategy in engineering *PVY*-resistant plants is the fact that the integrated shRNA sequence is not of viral origin nor it is translated into a protein. Moreover, the actual RNA transcript is almost undetectable because it gets cleaved quickly in small fragments through RNAi pathway. These two features limit the environmental risks of this strategy, such as trans-encapsidation or recombination of the transgene with an incoming virus.

Author details

Bushra Tabassum, Idrees Ahmad Nasir, Usman Aslam and Tayyab Husnain
National Centre of Excellence in Molecular Biology (CEMB), University of the Punjab, Lahore, Pakistan

Abbreviations

RNAi (RNA interference); *PVY* (Potato virus Y); *SCMV* (Sugarcane mosaic virus); PTGS (Post Transcriptional Gene Silencing); siRNA (small interfering RNA); shRNA (short hairpin

RNA); ssRNA (single stranded RNA); dsRNA (double stranded RNA); RdRp (RNA dependent RNA polymerase); hpRNA (hairpin RNA).

9. References

[1] Pakistan economic Survey (2010-2011) Government of Pakistan; Ministry of finance. http://www.finance.gov.pk/survey_0910.html.

[1] Beachy RN (1997) Mechanisms and applications of pathogen-derived resistance in transgenic plants. Curr. Opin. Biotechnol. 8:215–220.

[2] Briddon RW, Markham PJ (2000) Cotton leaf curl virus disease. Virus Res. 71:151-159.

[3] Novy RG, Nasruddin A, Ragsdale DW, Radcliffe EB (2002) Genetic Resistances to Potato Leafroll Virus, Potato Virus Y, and Green Peach Aphid in Progeny of *Solanum etuberosum*. American Journal of Potato Research 79(1): 9-18.

[4] Anonymous (1997) Sugar and sweetener situation and outlook yearbook. US department of Agriculture Washington DC.

[5] Hema M, Joseph J, Gopinath K, Sreenivasulu P, Savithri HS (1999) Molecular characterization and interviral relationships of a flexuous filamentous virus causing mosaic disease of sugarcane (*Saccharum officinarum* L.) in India. Arch. Virol. 144: 479–490.

[6] Koike T, Martin DP, Johnson EM Jr (1989) Role of calcium channels in the ability of membrane depolarization to prevent neuronal drath induced by trophic factor deprivation: evidence that levels of internal calcium determine NGF dependence of sympathetic ganglion cells. Proc. Natl. Acad. Sci. 86:6421-25.

[7] Akhtar KP, Ryu KH, Saleem MY, Asghar M, Jamil FF, Haq MA, Khan IA (2008) Occurrence of Cucumber mosaic virus Subgroup IA in tomato in Pakistan. J. Plant Dis. Protect. 115: 2-3.

[8] Solomon-Blackburn RM, Barker H (2001) Breeding virus resistant potatoes (Solanum tuberosum): a review of traditional and molecular approaches. Heredity 86:17-35.

[9] Cooper JI, Jones AT (1983) Responses of plants to viruses, proposals for the use of terms. Phytopathology 73:127-128.

[10] Walkey DGA (1991) Applied plant virology. 2nd edition. Chapman and Hall India.

[11] Wesley SV, Helliwell CA, Smith NA, Wang MB, Rouse DT, Liu Q, et al (2001) Construct design for efficient, effective and high throughput gene silencing in plants. The Plant Journal 27(6): 581-590.

[12] Beachy RN, Loesch-Fries S, Tumer NE (1990) Coat protein mediated resistance against virus infection. Annu. Rev. Phytopathol. 28:451-74.

[13] Gadani F, Mansky LM, Medici R, Miller WA, Hill JH (1990) Genetic engineering of plants for virus resistance. Arch. Virol. 115:1-21.

[14] Hull R, Davies JW (1992) Approaches to nonconventional control of plant virus diseases. Crit. Rev. Plant Sci. 11:17-33.

[15] Yusibov V, Loesch-Fries LS (1995) High-affinity RNA-binding domains of alfalfa mosaic virus coat protein are not required for coat protein-mediated resistance. Proc. Natl. Acad. Sci. USA 92:1–5.

[16] Carrington JC, Kasschau KD, Mahajan SK, Schaad MC (1996) Cell-to-cell and long-distance transport of viruses in plants. Plant Cell 8:1669-l681.

[17] Carr JP, Marsh LE, Lomonossoff GP, Sekiya ME, Zaitlin M (1992) Resistance to tobacco mosaic virus induced by the 54-kDa gene sequence requires expression of the 54-kDa protein. Mol. Plant–Microbe Interact. 5: 397–404.

[18] Golemboski DB, Lomonossoff GP, Zaitlin M (1990) Plants transformed with a tobacco mosaic virus nonstructural gene sequence are resistant to the virus. Proc. Natl. Acad. Sci. USA 87:6311–6315.

[19] Lomonossoff GP. Pathogen-derived resistance to plant viruses. Annu. Rev. Phytopathol. 33:323-343.

[20] Baulcombe DC. RNA as a target and an initiator of post-transcriptional gene silencing in transgenic plants. Plant Molecular Biology 32(1-2): 79-88.

[21] Farinelli L, Malnoe P, Collet GF (1992) Heterologous encapsidation of potato virus Y strain^O (PVY-^{O}) with the transgene protein of PVY strain (PVY-^{N}) in *Solanum tuberosum* cv. Bintje. Biotechnology 10:1020–1025.

[22] Kollar A, Thole V, Dalmay T, Salamon P, Balazs E (1993) Efficient pathogen-derived resistance induced by integrated potato virus Y coat protein gene in tobacco. Biochemie 75:623–629.

[23] Malnoe P, Farinelli L, Collet G, Reust W (1994) Small-scale field tests with transgenic potato, cv. Bintje, to test the resistance to primary and secondary infections with potato virus Y. Plant Mol. Biol. 25: 963–975.

[24] Smith HA, Powers H, Swaney S, Brown C, Dougherty WG (1995) Transgenic potato virus Y resistance in potato: evidence for anRNA-mediated cellular response. Phytopathology 85:864–870

[25] Maki-Valkama T, Pehu T, Santala A, Valkonen JP, Koivu K, Lehto K, Pehu E (2000) High level of resistance to potato virus Y expressing P1 sequence in antisense orientation in transgenic potato. Molecular Breeding 6: 95–104.

[26] Gargouri-Bouzid R, Jaoua L, Mansour R B, Hathat Y, Ayadi M, Ellouz R (2005) PVY resistant transgenic potato plants (cv. Claustar) expressing the viral coat protein. J. Plant Biotechnol. 3:1–5

[27] Hassairi A, Masmoudi, K, Albouy J, Robaglia C, Jjullien M, Ellouz R (1998) Transformation of two potato cultivars `Spunta' and `Claustar' (Solanum tuberosum) with lettuce mosaic virus coat protein gene and heterologous immunity to potato virus Y. Plant Sci. Limerick. 136: 31-42.

[28] Kaniewski W, Lawson G, Sammons B, Haley L, Hart J, Delan-nay X, Tumer NE (1990) Field resistance of transgenic Russet Burbank potato to effects of infection by potato virus X and potato virus Y. Biotechnology 8: 750–754.

[29] Lawson G, Kaniewski W, Haley L, Rozman R, Newell C, Sanders P, Tumer NE (1990) Engineering resistance to mixed virus infection in a commercial potato cultivar: resistance to potato virus X and potato virus Y in transgenic Russet Burbank. Biotechnology 8: 1277–134.

[30] Okamoto D, Nielsen SVS, Albrechtsen M, Borkhardt B (1996) General resistance against potato virus Y introduced into a commercial potato cultivar by genetic transformation with PVYN coat protein gene. Potato Research 39:271-82.

[31] Pehu TM, Maki-Valkama TK, Valkonen JPT, Koivu KT, Lehto KM, Pehu EP (1995) Potato plants transformed with a potato virus Y P1 gene sequence are resistant to PVY°. American Potato Journal 72: 523-532.

[32] Tacke E, Salamini F, Rohde W (1996) Genetic engineering of potato for board-spectrum protection against virus infection. Nature Biotechnology 19:1597-1601.

[33] Bejarano ER, Lichtenstein CP (1992) Prospects for engineering virus resistance in plants with antisense RNA. Trends Biotechnol. 10:383–388.

[34] Culver JN (1995) Molecular strategies to develop virus-resistant plants. CRC Press, Inc.

[35] Chi-Ham CL, Clark KL, Bennett AB (2010) The intellectual property landscape for gene suppression technologies in plants. Nat. Biotechnol. 28(1):32-36.

[36] Stam M, de Bruin R, Kenter S, van der Hoorn RAL, van Blokland R, Mol JNM, et al (1997) Post-transcriptional silencing of chalcone synthase in petunia by inverted transgene repeats. Plant J.12:63–82.

[37] Luff B, Pawlowski L, Bender J (1999) An inverted repeat triggers cytosine methylation of identical sequences in Arabidopsis. Mol. Cell. 3:505-511.

[38] Mette MF, van der Winden J, Matzke MA, Matzke AJ (1999) Production of aberrant promoter transcripts contributes to methylation and silencing of unlinked homologous promoters in trans. EMBO J. 18: 241–248.

[39] Pelissier T, Thalmeir S, Kempe D, Sanger HL, Wassenegger M (1999) Heavy de novo methylation at symmetrical and non-symmetrical sites is a hallmark of RNA-directed DNA methylation. Nucleic Acids Res. 27:1625-1634

[40] Wassenegger M (2000) RNA-directed DNA methylation. Plant Mol. Biol. 43: 203–220.

[41] Jones AL, Thomas CL, Maule AJ (1998) De novo methylation and co-suppression induced by a cytoplasmically replicating plant RNA virus. EMBO J. 17: 6385–6393

[42] Fire A, Xu S, Montgomery M, Kostas S, Driver S, Mello C (1998) Potent and specific genetic interference by double-stranded RNA in *Caenorhabditis elegans*. Nature 391(6669):806–811.

[43] Kasschau KD, Carrington JC (1998) A counter defensive strategy of plant viruses: suppression of posttranscriptional gene silencing. Cell 95:461-470.

[44] Anandalakshmi R, Pruss GJ, Ge X, Marathe R, Mallory AC, Smith TH, et al (1998) A viral suppressor of gene silencing in plants. Proc. Natl. Acad. Sci. USA 95:13079-84.

[45] Brigneti G, Voinnet O, Li WX, Ji LH, Ding SW, Baulcombe DC (1998) Viral pathogenicity determinants are suppressors of transgene silencing in *Nicotiana benthamiana*. EMBO J. 17:6739–6746.

[46] Bartel DP (2004) MicroRNAs: genomics, biogenesis, mechanism, and function. Cell. 116:281–97.

[47] Lecellier CH, Voinnet O (2004) RNA silencing: no mercy for viruses? Immunol. Rev. 198:285–303.

[48] Vastenhouw NL, Plasterk RH (2004) RNAi protects the *Caenorhabditis elegans* germline against transposition. Trends Genet. 20:314–319.

[49] Baulcombe DC (2004) RNA silencing in plants. Nature. 431(7006):356-63.

[50] Tang G, Reinhart BJ, Bartel DP, Zamore PD (2003) A biochemical framework for RNA silencing in plants. Genes Dev. 17:49–63.

[51] Hamilton A, Voinnet O, Chappell L, Baulcombe D (2002) Two classes of short interfering RNA in RNA silencing. EMBO J. 21:4671–4679.

[52] Tijsterman M, Plasterk RH (2004) Dicers at RISC; the mechanism of RNAi. Cell. 117:1–3.

[53] Gregory RI, Chendrimada TP, Cooch N, Shiekhattar R (2005) Human RISC Couples MicroRNA Biogenesis and Posttranscriptional Gene Silencing. Cell. 123:631–640.

[54] Schwarz DS, Hutvagner G, Du T, Xu Z, Aronin N, Zamore PD (2003) Asymmetry in the assembly of the RNAi enzyme complex. Cell. 115:199–208.

[55] Li J, Yang Z, Yu B, Liu J, Chen X (2005) Methylation protects miRNAs and siRNAs from a 3'-end uridylation activity in Arabidopsis. Curr. Biol.15:1501-1507.

[56] Fagard M, Vaucheret H (2000) Systemic silencing signal(s). Plant Mol. Biol. 43(2-3):285-93.

[57] Palauqui JC, Elmayan T, Pollien JM, Vaucheret H (1997) Systemic acquired silencing: transgene-specific post-transcriptional silencing is transmitted by grafting from silenced stocks to non-silenced scions. EMBO J. 15: 4738–4745.

[58] Voinnet O, Baulcombe DC (1997) Systemic signalling in gene silencing. Nature. 389(6651):553.

[59] Voinnet O, Vain P, Angell S, Baulcombe DC (1998) Systemic spread of sequence-specific transgene RNA degradation is initiated by localised introduction of ectopic promoterless DNA. Cell. 95:177–187.

[60] Winston WM, Molodowitch C, Hunter CP (2002) Systemic RNAi in C. elegans requires the putative transmembrane protein SID-1. Science. 295: 2456–2459.

[61] Palauqui JC, Balzergue S (1999) Activation of systemic acquired silencing by localised introduction of DNA. Curr. Biol. 9: 59–66.

[62] Mlotshwa S, Voinnet O, Mette MF, Matzke M, Vaucheret H, Ding SW, Pruss G, Vance VB (2002) RNA silencing and the mobile silencing signal. Plant Cell 14:289–301.

[63] Hamilton AJ, Baulcombe DC (1999) A species of small antisense RNA in post-transcriptional gene silencing in plants. Science 286:950-952.

[64] Hamilton A, Voinnet O, Chappell L, Baulcombe D (2002) Two classes of short interfering RNA in RNA silencing. EMBO J. 21:4671–4679.

[65] Plasterk RHA (2002) RNA silencing: the genomes immune system. Science 296:1263-1265.

[66] Bruening G (1998) Plant gene silencing regularized. Proc. Nat. Acad. Sci. 95:13349-13351.

[67] Waterhouse PM, Graham MW, Wang MB (1998) Virus resistance and gene silencing in plants can be induced by simultaneous expression of sense and antisense RNA. Proc. Natl. Acad. Sci. 95:13959–13964.

[68] Paddison PJ, Caudy AA, Bernstein E, Hannon GJ, Conklin DS (2002) Short hairpin RNAs (shRNAs) induce sequence-specific silencing in mammalian cells. Genes Dev. 16: 948–58.

[69] Horiguchi G (2004) RNA silencing in plants: a shortcut to functional analysis. Differentiation. 72:65-73.

[70] Watson JM, Fusaro AF, Wang M, Waterhouse PM (2005) RNA silencing platforms in plants. FEBS Lett. 579:5982–5987.

[71] Hirai S, Oka SI, Adachi E, Kodama H (2007) The effects of spacer sequences on silencing efficiency of plant RNAi vectors. Plant Cell Reports. 26(5):651-659.

[72] Meyer S, Nowak K, Sharma VK, Schulzer J, Mendel RR, Hansch R (2004) vector for RNAi technology in Poplar. Plant Biol. (Stuttg) 6. 100-103.

[73] Miki D, Shimamoto K (2004) Simple RNAi Vectors for St in Rice. Plant Cell Physiol. 45(4):490–495.

[74] Prins M, Laimer M, Noris E, Schubert J, Wassenegger M, Tepfer M (2008) Strategies for antiviral resistance in transgenic plants. Molecular plant pathology. 9(1):73–83.

[75] Missiou A, Kalantidis K, Boutla A, Tzortzakaki S, Tabler M, Tsagris M (2004) Generation of transgenic potato plants highly resistant to potato virus Y (PVY) through RNA silencing. Molecular Breeding 14: 185–197.

[76] Smith HA, Powers H, Swaney S, Brown C, Dougherty WG (1995) Transgenic potato virus Y resistance in potato: evidence for anRNA-mediated cellular response. Phytopathology 85:864–870.

[77] Tabassum B, Nasir IA, Husnain T (2011) Potato virus Y mRNA expression knockdown mediated by siRNAs in cultured mammalian cell line. Virol Sin. 26(2):105-13.

[78] Tabassum B, Nasir IA, Aslam U, Husnain T (2012) PhD dissertation. CEMB, University of the Punjab Lahore, Pakistan.

Chapter 7

Repetitive DNA: A Tool to Explore Animal Genomes/Transcriptomes

Deepali Pathak and Sher Ali

Additional information is available at the end of the chapter

1. Introduction

The analysis of genetic diversity and relatedness within and between the different species and populations has been a major theme of research for many biologists. With the availability of whole-genome sequencing for an increasing number of species, focus has been shifted to the development of molecular markers based on DNA or protein polymorphism. DNA sequences originate and undergo evolutionary metamorphoses' and thus may be used as powerful genetic markers to characterize genomes of wide range of species. This type of analysis is called fingerprinting, profiling or genotyping. DNA profiling based on typing individuals using highly variable minisatellites in the human genome was first developed by Jeffreys et al (1985). He demonstrated short repeat sequences tandemly arranged within the gene(s) and each organism has a unique pattern of the arrangement of these minisatellites, the only exception being multiple individuals from a single zygote (e.g. identical twins). DNA fingerprinting technique was notably used to help solve crimes and determine paternity. In addition, with the advances in Molecular biology techniques, isolation of genes tagged with minisatellites has become the most powerful tool for genome analysis.

The term "repetitive sequences" (repeats, DNA repeats, repetitive DNA) refers to DNA fragments that are present in multiple copies in the genome. These sequences exhibit a high degree of polymorphism due to variation in the number of their repeat units caused by mutations involving several mechanisms (Tautz, 1989). This hypervariability among related and unrelated organisms makes them excellent markers for mapping, characterization of the genomes, genotype phenotype correlation, marker assisted selection of the crop plants, molecular ecology and diversity related studies. The nature of repeats provides ample working flexibilities over the other marker systems. This is because: (i) short tandem repetitive (STR) sequences are evenly distributed all over the genome (ii), are often conserved between closely related species (iii) and are co-dominant. With these innate attributes, very small quantities of DNA can be used for simultaneous detection of the alleles tagged with STR employing minisatellite associated sequence amplification (MASA).

Current data base has the information on the genomes of various livestock species, like cattle, sheep, goat, pig, horse, chicken, Silkworm, Honey Bee, Rabbit, Dog, cat and duck (Georges and Andersson, 1996). However complete sequence analysis of several important species such as Yak, Banteng, Zebu, Donkey, Goose, Turkey, Camel and Water buffalo are still underway. Efforts are required to characterize genes controlling important traits in order to produce genetically healthy breeds and segregate superior germplasm wherever possible. Water buffalo, *Bubalus bubalis* is important domestic animal worldwide having immense potential in agriculture, dairy and meat industries. We have studied several repeat loci in buffalo genome using Restriction Fragment Length Polymorphism and characterized a number of important genes employing Minisatellite Associated Sequence Amplification (MASA).

MASA forms a rich basis of functional and comparative genomics contributing towards the understanding of genome organization, gene expression and development of molecular synteny. This approach also enables characterization of the same genes across the individuals within a species and amongst the individuals between the species. Thus, information about the organization of gene, its expressional, mutational and phylogenetic status, chromosomal location and genetic variations across the genomes maximize the chances of narrowing the search of possible genetic markers.

In this chapter, we discuss overall organization of the repetitive sequences, their origin, distribution, application in genome analysis and implications. In addition, use of repetitive sequences in bubaline genome mining is highlighted elucidating the potential of functional and comparative genomics. Thus, organizational variation and expressional profile of a single gene originating from a specific tissue may be studied in many ways to meet the varying requirements of biology.

2. Organization of repetitive sequences

The mammals have approximately 3 billion base pairs per haploid genome harboring about 20,000-25000 genes. A minor part of the genome (5-10%) is coding sequences (International Human Genome Sequencing Consortium, 2004; Hochgeschwender and Brennan, 1991) and the remaining part is non-coding representing repetitive DNA (Bromham, 2002). Comparison of the genome size of different eukaryotes shows that the amount of non-coding DNA is highly variable and constitutes 30% to about 99% of the total genome (Elgar and Vavouri, 2008; Cavalier-Smith, 1985). The non-coding repetitive sequences are dynamic elements, which reshape their host's genome by generating rearrangements, shuffling of genes and modulating pattern of expression. This dynamism of repeats leads to evolutionary divergence that can be used in species identification, phylogenetic inference and for studying process of sporadic mutations and natural selection. These repetitive sequences are mainly composed of interspersed and tandem repeats (Slamovits and Rossi,

2002). The later includes satellite, minisatellites and microsatellites. Satellite DNAs are predominantly associated with centromeric heterochromatin and the same is being increasingly utilized as a versatile tool for genome analysis, genetic mapping and for understanding chromosomal organization. On the other hand minisatellite and microsatellites are dispersed throughout the genome and are highly polymorphic in all populations studied. This arrangement has led to their extensive use as genetic markers for fingerprinting, genotyping, and for forensic analysis in human system. Based on their arrangements, repetitive DNA sequences are classed into two types (Figure 1).

2.1. Highly repetitive sequences

These are are short sequences (5 to10 bp) amounting 10% of the genome and repeated a number of times, usually occurring as tandem repeats (present in approximately 10^6 copies per haploid genome). However, they are not interspersed with different non-repetitive sequences. Usually, the sequence of each repeating unit is conserved. Most of the sequences in this class are located in the heterochromatin regions of the centromeres or telomeres of the chromosomes. Highly repetitive sequences interacting with specific proteins are involved in organizing chromosome pairing during meiosis and recombination.

2.1.1. Satellite DNA

These are represented by monomer sequences, usually less than 2000-bp long, tandemly reiterated up to 10^5 copies per haploid animals and located in the pericentromeric and or telomeric heterochromatic regions (Charlesworth et al 1994). Satellite DNA constitutes from 1 to 65% of the total DNA of numerous organisms, including that of animals, plants, and prokaryotes. The term "satellite" in the genetic sense was first coined by the Russian cytologist Sergius Navashin, in 1912, initially in Russian ("sputnik") and Latin (*satelle*), and was later translated to "satellite" (Battaglia, 1999). The more familiar usage of "satellite" relates to a small band of DNA with a density different (usually lower, because of a high AT-content) from the bulk of the genomic DNA, which are separated from the main band following CsCl centrifugation (Kit, 1961). Nucleotide changes and copy number variations fuel the process of their evolution within and across the species (Ugarkovic and Plohl, 2002). Satellite fraction(s), though not conserved evolutionarily (Bhatnagar et al 2004; Amor and Choo, 2002), are unique to a species and usually show similarity amongst related group of animals (Pathak et al 2011; Henikoff et al 2001; Ali and Gangadharan, 2000).

2.2. Moderately or dispersed repetitive sequences

These include short (150 to 300-bp) sequences or long ones (5-kbp) amounting about 40% and 1-2% of the total genome, respectively. These are dispersed throughout the euchromatin having 10^3-10^5 copies per haploid genome. These sequences are involved in the regulation of gene expression. In some cases, long dispersed repeats of 300 to 600-bp show homology with the retro viruses.

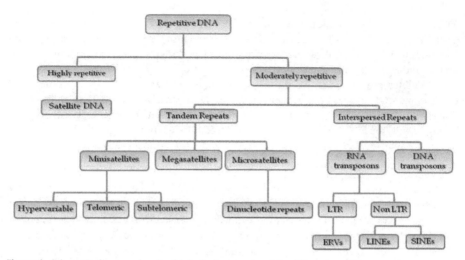

Figure 1. Schematic diagram showing biological categories of the different repetitive sequences.

On the basis of their mode of amplification, repetitive DNA sequences may be tandemly arranged or interspersed in the genome (Slamovits and Rossi, 2002).

2.2.1. Interspersed repetitive DNA

Interspersed repeat sequences scattered throughout the genome have arisen by transposition, having "ability to jump from one place to another in the genome" (Miller and Capy, 2004; Brown, 2002). Even though the individual units of interspersed repetitive non-coding DNA are not clustered, taken together they account for approximately 45% of the human genome. By the mechanism of their transposition, interspersed repeats are classified into two classes:

2.2.1.1. RNA transposons

RNA transposons also known as retroelements found in eukaryotic genome require reverse transcription for their activity. Based on their structural relationship, RNA transposons are divided into two general categories:

2.2.1.1.1. LTR elements

LTR includes retroviruses whose genomes are made up of RNA. They infect different types of vertebrates.

Endogenous retroviruses (ERVs)

These are retroviruses integrated into the vertebrate chromosomes and inherited from generation to generation as part of the host genome. Some are still active and might, at some stage in a cell's lifetime, direct synthesis of the exogenous viruses. However, majority of them are decayed relics and no longer have the capacity to form viruses (Patience et al 1997).

Retrotransposons are the biggest class of the transposons. An important characteristic of this type of transposable element is that they usually contain sequences with potential regulatory activity. They have features of non-vertebrate eukaryotic genomes (i.e. plants, fungi, invertebrates and microbial eukaryotes). These elements code for mRNA molecule which is processed and polyadenylated. Retrotransposons have very high copy numbers. In maize, these elements occupy half of the genome.

2.2.1.1.2. Non LTR elements

LINEs (Long Interspersed Nuclear Elements)

LINEs are several thousand base pairs in size and make up about 17% of the total human genome (Richard and Batzer, 2009). They contain reverse-transcriptase-like gene involved in retrotransposition process. Many LINEs also code for an endonuclease (e.g. RNase H). The most abundant LINE family is the 7-kbp, L1 repeat element having >500,000 copies and accounts for approximately 15% of the human genome (Lander et al 2001). Despite its abundance, no function of LINE 1 repeat is yet known. Initial studies on mouse have associated L1s in shaping the structure and expression of the transcriptomes (Han et al 2004; Han and Boeke, 2005).

SINEs (Short Interspersed Nuclear Elements)

SINEs are small elements, usually 100 to 500-bp in length, accounting for 11% of the human genome (Richard and Batzer 2009). SINEs do not have reverse transcriptase gene, instead they borrow reverse transcriptase enzymes from other retroelements. Well-known example of SINE in the human genome is *Alu* sequences (Capy et al 1998), which are 350 base pairs long, do not contain any coding sequences, and have over 1 million copies (Roy-Engel et al 2001)

2.2.1.2. DNA Transposons

DNA transposons do not require RNA intermediate and transpose in a direct DNA-to-DNA manner. In eukaryotes, DNA transposons are less common than retrotransposons, but they have a special place in genetics because a family of plant DNA transposons - the Ac/Ds elements of maize. There are two types of DNA transposons that both require enzymes coded by genes within the transposons.

2.2.2. Tandem repeats

Tandem repeats consists of repeat arrays of two to several thousand-sequence units arranged in a head to tail fashion. Tandem repeats may be further classified according to the length and copy number of the basic repeat units as well as its genomic localization.

2.2.2.1. Mega satellite DNA

These are characterized by tandemly repeated DNA in which the repeat unit is approximately 50-400 times, producing blocks that can be hundreds of kilobases long. Some mega satellites are composed of coding repeats. For example: RNA genes, and the deubiquitinating enzyme gene USP17.

2.2.2.2. Minisatellite DNA

This comprises tandem copies of repeats that are 6-100 nucleotides in length (Tautz, D. 1993). Alec Jeffrey's first described minisatellites in 1985, from the non-coding (intron) regions of the human myoglobin gene. Since then similar DNA structures have been reported in many organisms including bacteria (Skuce et al 2002), avian (Reed et al 1996), higher plants (Sykorová et al 2006; Durward et al 1995), protozoan (Feng et al 2011; Bishop et al 1998), and yeast (Kelly et al 2011; Haber and Louis, 1998) genomes. Comparison of the repeat units in classical minisatellites led to early notion of consensus or core sequences, which exhibit some behavioral similarities with the *Chi* sequences of λ phage (GCTGTGG). Also called as variable number of tandem repeats (VNTR) (Brown 2002), majority of the minisatellites are GC rich, with a strong strand asymmetry. Often minisatellites form families of related sequences that occur at many hundred loci in the nuclear genome. In human genome, number of minisatellite loci is estimated to be approximately 3000 and each locus contains a distinctive repeat unit with respect to size and sequence content. The degree of repetition ranges from two to several hundreds. Repeat unit within a minisatellite usually display small variations in sequence. Minisatellite mutations usually consist of gains or losses of one or more repeat units. Such mutations at hypervariable minisatellite loci are up to 1000 times more common than mutations in protein coding genes (Debrauwere et al 1997).

In the humans, majority of minisatellites are clustered near sub-telomeric ends of the chromosomes limiting their usefulness for extensive gene mapping (Lopes et al 2006; Royle et al 1988), but there are examples of interstitial locations (alpha globin gene cluster (Proudfoot et al 1982) and type II collagen gene (Stoker et al 1985). Minisatellites of other species, such as mice or bovine (Georges et al 1991), are not always preferentially clustered at chromosomal termini as in the human genome, but are distributed along the entire length of chromosomes. Unlike microsatellites, which usually alter during the DNA synthesis stage of the mitotic cell cycle, minisatellites alter during meiosis, undergoing changes in overall length and repeat composition (Jarman and Wells, 1989; Jeffrey's et al 1998). Minisatellite tracts have proven very useful for genomic mapping (Legendre et al 2007; Jeffrey's et al 1985) and linkage studies (Nakamura et al 1987). Examples of human minisatellite used for fingerprinting include consensus sequence of 33.6, 33.15 repeat loci. List of other minisatellite sequences according to Ali and Wallace, (1988) are mentioned in Table 1.

2.2.2.2.1. Telomeric repeats

These are composed of multiple repeats of short sequence elements (typically 5 to 8-bp in length, with a GT-rich strand oriented 5' to 3' toward the end of the chromosome) and range in length from a few repeat units to >10-kbp. Long simple sequence tandem repeats of interstitial TTAGGG arrays form a three-dimensional nuclear network of poorly transcribed domains, which involve gene silencing by repositioning. This network, as well as clusters of retroelements properly positioned in the nucleus, form unique lineage-specific structures that affect gene expression (Tomilin, 2008). The repeated sequence (TTAGGG)$_n$ is found at telomeres in all vertebrates, certain slime molds, and trypanosomes; (TTGGGG)$_n$ and

(TTTGGGG)$_n$ are found in the ciliated protozoan *Tetrahymena* and *Oxytricha* species, respectively; and (TG$_{1-3}$)$_n$ is found in the yeast *Saccharomyces cerevisiae*. In organisms whose telomeres have been examined in detail, the GT strand extends 12 to 16 nucleotides (two repeats) beyond the complementary C-rich strand. The unique structure of telomere is involved in the maintenance of the integrity of the chromosome ends.

S.No.	Probe name	Sequence 5'-3'	Total length	Length repeat unit	No. Repeat units
1	0-33.6-22	(CCTCCAGCCCT)$_2$	22	11	2
2	0-33.6-37	GCCCTTCCTCCGGAGCCCTCCTCCAGCC CTTCCTCCA	37	37	1
3	0-33.15-32	(CACCTCTCCACCTGCC)$_2$	32	16	2
4	0-33.15-80	(CACCTCTCCACCTGCC)$_5$	80	16	5
5	0-AY-29	GAGGARYAGAAAGGYGRGYRVTGTGGG CGC	29	37	1
6	0-YN-124	TCCTGAACAACCCCACTGTACTTCCCA	27	31	1
7	0-33.1	GTGCCTGCTTCCCTTCCCTCTCTTGTC	27	62	1
8	0-34BHI	CCTGCTCCGCTCACGTGGCCCACGCAC	27	?	1
9	0-CCR-26	(CCR)$_8$CC	26	3	8.67
10	0-CCA-26	(CCA)$_8$CC	26	3	8.67
11	0-H-Ras	CACTCCCCCTTCTCTCCAGGGGACGCCA	28	28	1
12	0-GACA-16	(GACA)$_4$	16	4	4
13	0-GACA-24	(GACA)$_6$	24	4	6

Table 1. Sequences and hybridization characteristics of the oligonucleotides probes (Ali and Wallace, 1988)

2.2.2.2.2. Subtelomeric repeats

Sub-telomeric Repeats are the classes of repetitive sequences that are interspersed within the last 500,000 bases of non-repetitive DNA located adjacent to the telomere. Some sequences are chromosome specific whereas others seem to be present near the ends of all the human chromosomes (Norman, 2001).

2.2.2.3. Microsatellite /Short Sequence Repeats (SSRs)

Tandem repeats are made up of usually, di-, tri-, or tetranucleotide units (1-6 bps), were earlier called simple sequences (Tautz and Renz, 1984). Later, this class of DNA was coined as microsatellites by Tautz 1989. Microsatellites or simple sequence repeats (SSRs) are ubiquitously interspersed in coding and non-coding regions of the eukaryotic and prokaryotic genomes (Gur-Arie et al 2000; Toth et al 2000). All the SSRs taken together occupy about 3% of the human genome in which they are widely dispersed and associated with many genes (Subramanian et al 2003). The significance of specific microsatellite in different regions has not been completely understood. However, some microsatellites occurring in flanking regions of coding sequences are believed to play significant roles in regulation of gene expression by forming various DNA secondary structures and offering a mechanism of unwinding (Catasti et al 1999). The variation of length and unit type of simple repeats in upstream activation sequences might influence transcriptional activity (Kim and Mullet, 1995; Epplen et al 1996; Martienssen and Colot, 2001; Zhang et al 2004), and affect interaction with different regulatory proteins during translation (Lue et al 1989).

Microsatellites are usually characterized by low degree of repetition at a particular locus. However, these elements containing identical motifs may be found at many thousand genomic loci. When the occurrence of SSRs in different functional genome regions is considered, it turned out that most of them show much higher density in non-coding regions. Exceptions to the rule are trimers and hexamers that are nearly two times more prevalent in exons compared to introns and intergenic regions. Their high frequency in coding regions may be explained by the fact that they do not change the reading frames and gene coding properties, thus, are much better tolerated than other SSRs. Their positive selection in exons suggests some functions for these repeats.

The high mutation rate of these repeats and their frequent length polymorphism suggest that they may be involved in the regulation of gene expression thus leaving quantitative effects on the phenotype. Few examples of repeat units used for fingerprinting and transcriptome analysis includes (GATA/GACA)n, CA, (AT)n, (GAA)n, (TCC)n, (GGAT)n, (GGCA)n, and (TTAGGG)n.

3. Evolution and inter-species variation of repeat sequences

Several mechanisms have been proposed for their evolution, such as stand slippage during replication, base misalignment and unequal cross over between homologous chromosomes during meiosis, sister chromatid exchanges or even insertion of the viral genome (Barros, 2008; Jeffrey's et al 1985; Tautz, 1989).

Microsatellites tend to be highly polymorphic, suggesting a 'stepwise mutation' model in which most variations are introduced by replication slippage, changing the array length by only one or two repeats at a time, but also with occasional larger 'jumps' in size at much lower frequency. Minisatellites, evolve more readily by larger-scale mechanisms such as unequal exchanges. For all classes, there appears to be a general bias towards increase in array length through evolutionary time. Highly repetitive DNA tends to accumulate only in regions of low recombination such as centromeres and telomeres, where recombination is suppressed, while repeats occurring in euchromatin are much more susceptible to crossing-over and tend to be more variable in copy number relative to their array length.

As mentioned above, mechanism of loss or gain of repeat by unequal cross over and gene conversation can lead to molecular drive of any given variant in a sexually dimorphic population. During the evolution of repetitive elements by unequal cross over, some variants will be lost whereas others will increase in frequency, eventually replacing all others. These evolutionary changes leads to homogeneity in the repeats of an array within a species and heterogeneity in the units of the corresponding array in different species, giving rise to inter- species variations (Harris and Wright, 1995). This phenomenon however is affected by overall male female ratios, population size and possibility of infusion of newer genetic materials in a given gene pool and allele fixation involving evolutionary incubation time.

4. Functional significance of repetitive sequences

With respect to functional roles of these sequences, uncertainty persisted for a long time and it was largely believed that they represent detritus part of the genome (Ohno, 1972). However, recent studies have shown repeat elements influencing the structure, function, and evolution of the chromosomes in the host genomes (Sinden, 1999; Dey and Rath, 2005; Tang, 2011). Their association with the promoters and coding regions of the genes has made them very attractive objects of the study. Transcription, mRNA processing, translation, folding, stability and aggregation rates, as well as gross morphology have been found to be incrementally affected by the alterations in the tracts of tandem repeats (Fondon and Garner, 2004; Vinces, 2009). The human genome provides many instances of regulatory regions embedded in the remnants of repeat elements (Jordan et al. 2003) and studies have documented participation of repeat sequences in regulation of gene expression (Boeva et al., 2006). This suggests that the repeat elements play a major architectonic role in higher order of physical structuring of the genome (Shapiro and Sternberg, 2005; Vermaak et al 2009). More studies on repeat sequences will lead to an increased understanding on the functions and dysfunctions of the genomes.

5. Significance of repetitive sequences as marker

Primers based on VNTR provide an unprecedented opportunity to develop potential molecular markers for a particular species. Where a complete genome sequence is available for an organism, repeats may be annotated with their physical position on the genome. Markers may then be selected either for their location within a specific region of interest or for their even distribution across the regions. Where a full genome sequence is unavailable, location may be predicted through synteny using a sequenced genome or through previous mapping exercises. Alternatively, for a genome whose sequences are not known can still be analyzed employing primers from other species for gene amplification. A gene so amplified may then be localized onto the chromosomes employing FISH. Similar set of primer may be used to amplify cDNA of the species. This approach circumvents the need for screening the genomic library.

Furthermore, for species which exhibit low levels of polymorphism at repeat loci, candidate polymorphic loci may be predicted through mining large sequence datasets. The presence of short sequence repeat (SSR) polymorphisms within aligned sequences of different origin would be indicative of the level of polymorphism at that locus. These selection strategies could greatly reduce the time and cost associated with the development of repeat markers. Integration of this repetitive sequence data with genome databases would provide further benefits to genome researchers.

6. *Bubalus bubalis* genome

The water buffalo (*Bubalus bubalis*) population in the world is actually about 168 million head, of which 161 million can be found in Asia (95.83 percent); 3717 million are in Africa and Egypt (2.24 percent); 3.3 million (1.96 percent) in South America, 40 000 in Australia (0.02 percent); 500 000 in Europe (0.30 percent). Asian buffalo or Water buffalo is classified under the Genus: *Bubalus*, Species: *bubalis*. Asian buffalo includes two subspecies known as the River and Swamp types, the morphology and purposes of which are different so are the genetics. The River buffalo has 50 chromosomes of which five pairs are sub-metacentric, while 20 are acrocentric: the Swamp buffalo has 48 chromosomes, of which 19 pairs are metacentric. Swamp buffaloes are stocky animals with marshy land habitats. They are primarily used for draught power in paddy fields and haulage but are also used for meat and milk production. They produce a valuable milk yield of up to 600 kg milk per year, Swamp buffaloes are mostly found in South East Asian countries. A few animals can also be found in the northeastern states of India (Sethi, 2003). River buffaloes are generally large in size, with curled horns and are mainly found in India, Pakistan and in some countries of western Asia. They prefer to enter clear water, and are primarily used for milk meat and draught purposes. Each subspecies includes several breeds. Buffaloes are known to be better at converting poor-quality roughage into milk and meat. They are reported to have a 5 percent higher digestibility of crude fiber than high-yielding cows; and a 4-5 percent higher efficiency of utilization of metabolic energy for milk production (Mudgal, 1988).

India has about 97 million animals, which represents 92% of the world buffalo population. India possesses the best River milk breeds in Asia e.g. Murrah, Nili-Ravi, Surti Jaffarabadi, Mehsana, Kundi, Bhadavari and Nagpuri which originated from the north-western states of India (Sethi, 2003). However, despite the importance of buffalo to the economic and social fabric of the region, its population has been declining. There are many reasons for the decline of buffalo populations, foremost of which are: increased agricultural mechanization; increased urbanization, industrialization, and reforestation limiting paddy areas for buffaloes; growing buffalo slaughter rate to satisfy meat demands of a fast-growing population; poor reproductive performance; and lack of proper attention by policy makers and researchers. The low reproductive efficiency in female buffalo can be attributed to delayed puberty, higher age at calving, long postpartum anoestrus period, long calving interval, lack of overt sign of heat, and low conception rate. In addition, female buffaloes have few primordial follicles and a high rate of follicular atresia. Understanding potential quantitative trait loci associated with economically important traits will help in segregating genetically superior breeds.

7. Repetitive sequences as molecular markers in bovid genome

Based on repeat sequences, a number of probes with varying length and sequence complexities have been successfully used as genetic markers (Kapur et al 2003; Jobling and Tyler-Smith, 2003; Bashamboo and Ali, 2001; Amos et al 1991; Tourmente et al 1994; Ali et al 1986). Earlier conventional protein and biochemical markers were used for breeding program of bubaline species (Wilson and Strobeck, 1999). Subsequently, diallelic Restriction Fragment Length Polymorphism (RFLP) for the loci homologous to cattle (Blott et al 1999) were used. However due to low levels of polymorphism detected with these markers, their application remained limited. RFLP technology was followed by Random amplification of Polymorphic DNA (RAPD), followed by Amplified fragment length Polymorphism (AFLP) besides minisatellite markers. A series of synthetic oligonucleotide probes were developed as markers for genetic analysis and molecular systematics of Bubaline and related genomes. While probes based on repeat sequences are available there is no clear cut experimental approach that could assist identification and segregation of elite animals with superior QTL loci. This is because most of the physical and physiological attributes recognized to be the part of the elite animals, are controlled by several genes and it is extremely challenging to uncover all such genes implicated with superior germplasm. However marker based analysis would possibly bridge the gap and facilitate much-needed advance research to segregate genetically superior germplasm in the context of animal genetics in general and animal biotechnology in particular.

8. Restriction Fragment Length Polymorphism (RFLP)

The basic technique for detecting RFLPs involves the fragmentation of genomic DNA by a restriction enzyme. The resulting DNA fragments are then separated by length through a process known as agarose gel electrophoresis, and transferred to a membrane via the Southern blot procedure. Hybridization of the membrane to a labeled DNA probe then determines the size of the fragments, which are complementary to the probe. An RFLP occurs when the size of a detected fragment varies between individuals. Each fragment size is considered an allele, and can be used in genetic analysis. RFLP's are quick, simple and inexpensive ways to assay DNA sequence differences. It is the first DNA polymorphism to be widely used for genomic characterization, which detects variations ranging from gross rearrangements to single base changes. The polymorphisms are found by their effects on sites for restriction enzyme mediated cleavage of preparations of high molecular weight DNA. In buffalo, RFLP approach has been used to gain insight into organization and allele length variation of satellite fractions (Chattopadhyay et al 2001; Bhatnagar et al 2004). From our laboratory, *BamH*1 derived pDS5 and pDS4 and *Rsa*I derived pDp1-pDp4 were found to be conserved only in buffalo, cattle, goat, and sheep (Pathak et al 2006; 2011).

9. Minisatellite Associated Sequence Amplification (MASA)

MASA involves random amplification of genomic or cDNA with primers specific to minisatellites by PCR. MASA can be performed with a small quantity of target substrate.

The novel part of the current approach is that functional, structural and regulatory genes associated with minisatellites are accessed without screening the conventional cDNA library proving this be highly useful for such genome analysis where prior information is absent or inadequately available. The expression profile of genes based on MASA under normal and abnormal conditions is envisaged to be of great relevance for identification of event/stage specific mRNA transcripts. In the context of comparative genomics, mRNA transcripts commonly expressing in a large number of species may be segregated. Following this approach, genes with highest levels of expression in a given tissue may be easily identified and the information from different breeds of animals may be established. In addition, differential expression of genes accessed by MASA may be used to establish genotype phenotype correlation in the context of genetic diseases, cancer biology, stem cell research, tissue engineering, organ transplantation, animal cloning, characterization of genetic integrity of different cell lines and conducting translational research. Minisatellite sequences 33.6, 33.15 have been widely used to explore bubaline genome (Srivastava et al 2006; 2008; Pathak et al 2010). In addition microsatellite probes (2-6 base pairs) such as (AT)n, (CA)n, (GAA)n, (TCC)n, (GACA)n, (GATA)n, (GGAT)n, (GGCA)n and (TTAGGG) were used to analyze buffalo genome (Rawal et al 2012; kumar et al 2011). Following this approach, additional oligo primers based on VNTR loci may be used to undertake analysis of any desired species, cell lines, biopsied samples and cell lines.

10. Technical approaches and methodologies

We describe some of our works related to characterization of the buffalo genome. Further, in the context of functional and comparative genomics, DNA from across the species were also used. DNA was largely procured from the blood samples though in some cases, solid tissues were also used.

10.1. Collection of blood samples and isolation of genomic DNA

DNA was extracted from peripheral blood of buffalo *Bubalus bubalis*, goat *Cipra hircus* sheep *Ovis aries* tiger *Panthera tigris*, lion *Panthera leo*, humans *Homo sapiens*, langur *Presbytis entellus*, Indian rhinoceros *Rhinoceros unicornis*, fish *Hetropnustes fossilis*, bird *Columba livia*, baboon *Papio hamadryas*, pig *Sus scrofa*, rat *Rattus norvegicus*, jungle cat *Felis chaus*, bonnet monkey *Macaca radiate* and leopard *Panthera pardus*. Intactness of DNA was checked on 1% agarose gel and DNA was PCR amplified using bubaline derived β actin primers and visualized on UV transilluminator.

10.2. RNA isolation and synthesis of cDNA

Using buffalo as an experimental animal, total RNA was extracted from testis, kidney, liver, spleen, lung, heart, ovary, brain and sperm using TRIzol (Molecular Research Center, Inc., Cincinnati, OH) following manufacturer's instructions. To check the contamination of mRNA from the cells other than spermatozoa, RNA extractions from the sperms were tested by RT-PCR both for the *CDH1* (E-cadherin) and *CD45* (tyrosine phosphatase). Similarly, presence of DNA was ruled out by PCR using β-actin primers. Following this, approximately 10 μg of RNA from different tissues and spermatozoa was reverse transcribed into cDNA using commercially available high capacity cDNA RT kit (Applied Biosystems, USA). The success of cDNA synthesis was confirmed by PCR employing 35 cycles of amplification using buffalo derived β-actin primers.

10.3. Minisatellite Associated Sequence amplification (MASA)

Using oligo primer and cDNA from different tissues and spermatozoa, PCR amplifications were carried out. The reaction conditions involved 95°C denaturation for 5 min followed by 35 cycles each consisting denaturation at 95°C for 1 min, annealing at the optimal temperature for 1.5 min, extension of the primer at 72°C for 1 min and final extension at 72°C for 10 min. Approximately, 25 μl of amplified product was resolved on a 20-cm-long, 3% (w/v) agarose gel in 1× TBE buffer at a constant voltage. The distinct bands were sliced from the gel, purified and cloned into pGEMT-easy vector (Promega, USA). In water buffalo, *Bubalus bubalis* using cDNA from the spermatozoa and eight different somatic tissues and an oligo primer based on two units of consensus of 33.6 repeat loci (5' CCTCCAGCCCTCCTCCAGCCCT 3'), Minisatellite-associated sequence amplification (MASA) identified 29 mRNA transcripts (Figure 2).

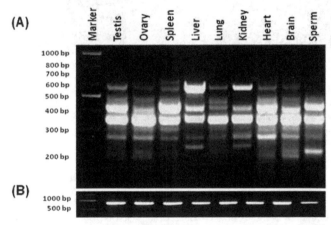

Figure 2. A representative agarose gel showing minisatellite associated sequence amplification (MASA) with cDNA from different somatic tissues and spermatozoa of buffalo as shown on top of the lanes in panel **(A)**. β-actin was used as an internal control **(B)**. M is the molecular marker given in base pairs (bp) (for details, see Pathak et al, 2010).

10.4. Restriction digestion of buffalo genomic DNA

Approximately, 4-5 μg of genomic DNA from buffalo, cattle, goat and sheep were subjected individually to restriction digestion using 4-5 units of *Bam*HI and *Rsa*1enzyme. The digested

DNA fragments were resolved on 0.8% agarose gel in 0.5X TBE for approximately 16-18 hours. In water buffalo, two distinct DNA bands of 1378 and 673 bp with Bam *HI* and four bands of 1331, 651, 603 and 339 base pairs were cut, gel purified (Figure 3). The eluted fragments were cloned and sequenced following standard protocol. For Southern hybridization, DNA was transferred onto Nylon membrane and immobilized by exposure to UV. Membranes were rinsed in 2X SSC, dried and UV cross- linked. Blots were hybridized at 60⁰C overnight with ^{32}P α-dCTP labeled recombinant plasmid (25 ng) using random priming method (rediprimeTM II kit, Amersham Pharmacia biotech, USA). Washing of the membranes was done using standard protocols and signals were recorded by exposure of the blot to X-ray film (Pathak 2006; 2011).

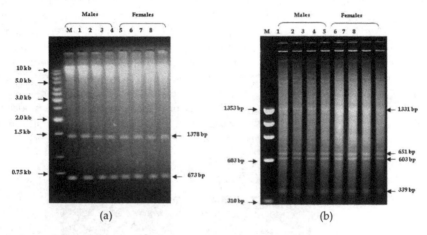

(a)   (b)

Figure 3. Agarose gel showing restriction digestion of buffalo *Bubalus bubalis,* genomic DNA with *Bam*HI **(A)** and *Rsa*I **(B)** enzymes. The two discernible bands 673 bp and 1378 bp with *Bam*HI digestion and four bands 1331, 651, 603 and 339 bp in *Rsa*I are highlighted. Molecular weight marker is given on the left in base-pair (bp). Since the patterns are not gender species, this suggests that the bands are originated from the autosomes (for details, Pathak et al 2006; 2011).

10.5. Copy number assessment and relative expression using Real Time PCR

Copy number of desired fragment was calculated based on absolute quantitation assay using SYBR Green dye and Sequence Detection System- 7500 (ABI, USA). The primers specific to fragments, respectively, were designed using Primer Express Software V2.0 (ABI). The standard curve was obtained using 10 folds dilution series of the recombinant plasmids ranging from 30, 00,000 to 30 copies taking 3.36 pg DNA per haploid genome of (assuming haploid genome of farm animals =3.3 pg, wt per base pair = 1.096×10^{-21} gm) as standards. The reactions were performed in triplicate using 96 well plates in a 25 µl reaction volume, each having 0.5 ng of buffalo genomic DNA and 50 nM of corresponding primers, employing conditions of 50⁰C for 2 min, 95⁰C for 10 min, followed by 40 cycles of 95⁰C for 10 sec and 60⁰C for 1 min. Real-time PCR analysis uncovered 1234 and 3420 copies of pDS5 and pDS4 fragments per the haploid genome and ~2 × 10⁴ copies of pDp1, ~ 3000 copies of pDp2 and pDp3 and ~ 1000 of pDp4 in buffalo, cattle, goat and sheep genomes (Figure 4), respectively. (Pathak et al 2006; 2011) The copy number assessment of these repeats in different known and nondescript breeds of buffalo may enable to establish a correlation, if any, towards the delineation of different breeds.

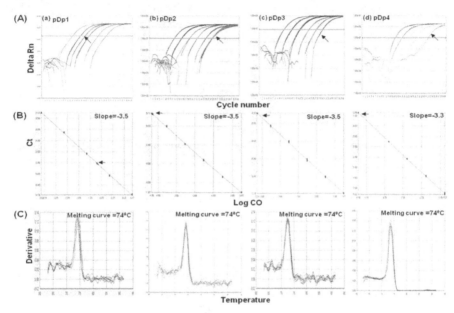

Figure 4. Standard curve based on 10 fold dilution series of pDp1, pDp2, pDp3, pDp4 and genomic DNA from buffalo, cattle, goat and sheep showing the amplification plot (a-d) panel (A), corresponding slopes of -3.3 to -3.5, panel (B) and a single dissociation peak, panel (C), substantiating maximum efficiency of the PCR reaction and high specificity of the primers with target DNA. Arrow indicates genomic DNA from buffalo, cattle, goat and sheep Buffalo, cattle, goat and sheep showed approximately same copy number for pDp1, pDp2, pDp3, pDp4 indicating their conservation across the bovid species (Pathak et al 2011).

Relative expression using Real Time PCR was carried out for the desired fragments with Sybr Green assay using cDNA from different tissues and spermatozoa. The primers were designed using Primer Express 2.0 (Applied Biosystems) software. The cyclic conditions were same as that used for copy number calculations. The reaction was performed following standard protocol (Sriavastava et al 2006). The specificity of each primer pair and the efficiency of the amplification were tested by assaying serial dilutions of the cDNA hybridized with oligonucleotides specific for target and normalization control (*GAPDH*). The difference in the Ct value between the target cDNA from different tissues and the control samples (the tissue showing least expression) was used for calculation. The expression level of the desired fragments was calculated using the formula: expression =

$(1+E)^{-\Delta Ct}$, where E is the efficiency of the PCR and ΔCt = difference in threshold cycle value between the test sample and endogenous control. To achieve the maximum (one) efficiency of the Real Time PCR, the amplicon size was kept small (70-150 bp) so that the expression level of the test gene remains $2^{-\Delta Ct}$. Each experiment was repeated three times to ensure consistency of the results. Maximum expression of pDS5 and pDS4 was seen in the spleen and liver, respectively. pDp1 showed maximum expression in lung, pDp2 and pDp3 both in Kidney, and pDp4 in ovary. Nine, 33.6 MASA amplified transcripts showed highest expression in spermatozoa and one each in liver and lung (Figure 5).

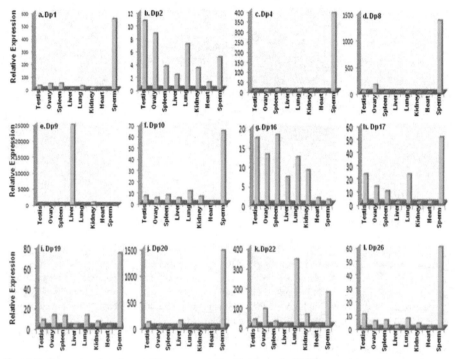

Figure 5. Expressional analyses of the representative 33.6 tagged mRNA transcripts. a-l represents different tissues, gonad, and spermatozoa. Note the maximum expression of some representative mRNA transcripts in the spermatozoa corresponding to Dp1, 4, 8, 10, 17,19, 20, and 26 shown in a, c, d, f, h, i, j, and l), respectively, and exclusive expression of Dp9 in liver (e). Bars represent relative expression of the transcript(s) in folds. Transcript IDs are mentioned on top left corner and tissues, below the panels (Pathak et al 2010)

These 9 transcripts in the spermatozoa, representing vital genes supports their involvement in sperm development and possibly overall testicular functions. In the context of animal biotechnology, such selective tissue specific expression profile is very important to segregate the genetically superior germplasm or any other physical and physiological attributes. This is true particularly in case of buffalo since this species has several breeds. Clearly, development of MASA mediated tissue specific transcript profiles is envisaged to go a long way in undertaking molecular characterization of not only the genome of buffalo but also those of other economically important animals.

10.6 Chromosome preparation and Fluorescence *In Situ* Hybridization (FISH)

When once mRNA transcripts are made available, it becomes feasible to fish out full length cDNA clones employing 3' and 5'RACE. This approach enables isolation of genes without screening the cDNA library thus circumventing many arduous steps. When clones representing genes are obtained, these can be used for conducting fluorescence hybridization onto the metaphase chromosomes. We have used some of the clones to successfully conduct FISH on buffalo chromosomes to localize the genes and also uncover the distribution of several species of repeat elements.

For chromosome culture, 2 ml of blood was drawn into heparinized vacutainer tubes with sterile syringe from buffalo, cattle, goat, sheep and human. To sterile tissue culture flask containing 5 ml RPMI- 1640, 20% fetal Bovine serum, 2% Phytohemagglutinin, PHA (2 mg/ml), 5 µl concavalin A (3 µg/ml), 2.5 µl mercaptoethanol (50 µM), 50 µl LPS (10 µg/ml), 2.2 ml Antibiotic/antimycotic (0.15mg/ml), 500 µl of blood was added and whole mixture was incubated for 72 hours in 5% CO_2 at 37^{0}C. After 70 hours, colcemid (10 µg/µl) was added in culture flasks to arrest cells in metaphase stage and cells were further incubated for 2 hours. The cells were then subjected to 0.56%KCL for 30 mins followed by fixative treatment (Methanol: glacial Acetic acid, 3:1). Few drops of cell suspension were dropped onto pre-cleaned chilled slide and blow-dried. The slides were Giemsa stained (Gibco, BRL) for 20 minutes, washed with PBS / distilled water and observed under microscope to record metaphases.

For probe preparation plasmids containing gene of interest were labeled with desired fluorochromes using Nick Translation Kit from Vysis, (Illinois, USA) following supplier's instructions. Hybridization was carried out in 20 µl volume containing 50% formamide, 10% Dextran sulphate, Cot 1 DNA and 2X SSC, pH 7 for 16 hours at 37^{0}C in a moist chamber. Post hybridization washes were done in 2X SSC at 37^{0}C (low stringent condition) and then at 60^{0}C in 0.1X SSC (under high stringent condition). Slides were counterstained with DAPI, screened under Olympus Fluorescence Microscope (BX51) and images were captured with Olympus U-CMAD-2 CCD camera. Chromosome mapping was done following the International System for Chromosome Nomenclature. The pDS5, representing the 1378-bp fragment, showed FISH signals in the centromeric region of acrocentric chromosomes only, whereas pDS4, corresponding to 673 bp, detected signals in the centromeric regions of all the chromosomes. *Rsa*I derived pDp1, pDp2 and pDp3 showed distribution of repeats to all across the buffalo chromosomes (Pathak et al 2006; 2011) (Figure 6).

Chromosomal mapping of *SARS2* gene, using bovine *SARS2* BAC probe localized (Figure 7) the genes to buffalo metaphase chromosome 18 (Pathak et al 2010).

(A)

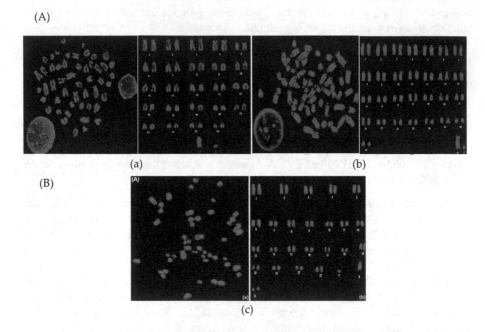

(a)   (b)

(B)

(c)

Figure 6. Chromosomal localization of pDS5 clone **(A)** on **(a)** buffalo and **(b)** cattle metaphase chromosomes. Note the absence of signal in all the bi-armed chromosomes. Fluorescence *in situ* hybridization (FISH) of pDp1 **(B)** clone on buffalo metaphase chromosomes **(c)**. Note the dispersed signals over the metaphase chromosomes. (See Pathak et al 2006; 2011).

11. Applications in animal biotechnology

With the availability of human genome sequence much emphases is given to sequence all the potential farm animals. Despite of importance as farm animal research data on the water buffalo is limited. Water buffalo breeders and farmers have been facing many challenges and problems, such as poor reproductive efficiency, sub-optimal production potential, higher than normal incidence of infertility, and lower rates of calf survival. Genome research has created a broad basis for promoting and utilizing gene technologies in many fields of livestock production. Genome biotechnology will provide a major opportunity to advance sustainable animal production systems of higher productivity through manipulating the variation within and between breeds to realize more rapid and better-targeted gains in breeding value. This type of research will also make it possible to distinguish molecular phenotypes and thus improve the use of genetic resources of domestic animals.

To date, researchers have identified several genes or DNA regions that are associated with traits of economic importance including reproduction, growth, lean body, fat quantity, meat quality, physical traits and disease resistance. For example, the fatty acid composition of dairy and beef food products, increased disease resistance (and thus increased animal welfare). Similarly, decreased methane emissions in cattle help to address the needs of consumers and society for sustainable and cost-effective food production. A number of gene and marker tests are now available commercially from genotyping service companies. Examples are CAST (meat quality), ESR and EPOR (litter size), FUT1 (E. coli disease resistance), HAL (halothane – meat quality, stress), IGF2 (carcase), MC4R (growth and fat), PRKAG3 (meat quality) and RN (meat quality) (Walters, 2011). Many scientists are using genomic information embedded on SNP50 BeadChip, a glass slide containing thousands of DNA markers, to determine disease-resistant genes in cattle, swine, sheep, poultry, fish and then selectively mating the animals in order to create disease resistant animals. Understanding all the expressed genes, their organization and mode of action in bubaline or any other farm animals will positively bridge the gap and facilitate the much needed growth of animal biotechnology.

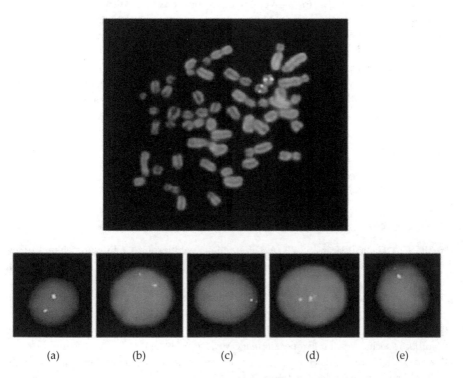

(a) (b) (c) (d) (e)

Figure 7. Localization of SARS2 gene on the representative interphase nuclei and metaphase chromosome 18 using FISH.SARS2 BAC probe showing signals on buffalo metaphase chromosome 18 (A) and interphase nuclei (a-e) (B). Note two signals in the interphase nuclei corresponding to those on the homologous chromosomes. (Pathak et al 2010).

12. Concluding remarks

Genetic improvement of animals warrants continuous and complex processes of sustained research employing cutting edge tools and techniques of modern biology and recombinant DNA technology. A much deeper and detailed understanding on a given species would eventually prove to be highly useful for possible manipulation of a desired genome. Improvement of domestic animal traits has been the foremost important task for animal breeding. In this pursuit, many techniques have been developed and tested. In recent years, advances in molecular genetics have introduced a new generation of molecular markers for the genetic improvement of the animals. However, utilization of marker-based information for genetic improvement depends on the choice and judicious use of an appropriate marker system for a given application. Selection of markers for different applications is influenced by the degree of polymorphism, reproducibility of the technique, speed of the experiments and cost involved.

As the situation stand now, for a given biological phenomenon where multiple genes are implicated, technical approaches need to be developed to segregate the entire possible genes specific to that phenomenon. A good example is the spermatogenesis that involves putatively close to about 400 plus genes. However, their clear cut involvement and characterization in any species has still not been achieved. When once, such information is made available, this would then provide much needed basis of functional and comparative genomics. Perhaps then, molecular delineation of the "so-called" elite animals or specific breed representing superior germplasm would become feasible.

Author details

Deepali Pathak and Sher Ali
National Institute of Immunology, Aruna Asaf Ali Marg, New Delhi, India

13. References

[1] Ali, S., & Gangadharan, S. (2000). Differential Evolution Of Coding And Non-Coding Sequences In Related Vertebrates: Implications In Probe Design. *Proc. Ind. Nat. Sci. Acad.* 664, 9-67.

[2] Ali, S., & Wallace, R. B. (1988). Intrinsic Polymorphism Of Variable Number Tandem Repeat Loci In The Human Genome. *Nucleic Acids Res.* 16, 8487-8496.

[3] Ali, S., Müller, C.R., & Epplen, J.T. (1986). DNA finger printing by oligonucleotide probes specific for simple repeats. *Hum Genet.* 74(3):239-243.

[4] Amor, D.J., & Choo, K. H. (2002). Neocentromeres: Role In Human Disease, Evolution, And Centromere Study. *Am. J. Hum. Genet.* 71, 695–714.

[5] Amos, W., & Hoelzel, A.R. (1991). Long-Term Preservation Of Whale Skin For DNA Analysis. *Rep. Int. Whal. Commn.* 13, 99–103.

[6] Barros, P., Blanco, M.G., Boán, F., & Gómez-Márquez, J. (2008). Evolution of a complex minisatellite DNA sequence. *Mol Phylogenet Evol.* 49(2):488-494.

[7] Bashamboo, A., & Ali, S. (2001). Minisatellite Associated Sequence Amplification (Masa) Of The Hypervariable Repeat Marker 33.15 Reveals A Male Specific Band In Humans. *Mol. Cell. Probes* 15, 89-92.

[8] Battaglia, E. (1999). The Chromosome Satellite (Navashin's "Sputnik" Or Satelles): A-345-Terminological Comment. *Acta Biologica Cracoviensia, Series Botanica* 41, 15-18.

[9] Bhatnagar, S., Bashamboo, A., Chattopadhyay, M., Gangadharan, S., & Ali, S. (2004). A 1.3 kb satellite DNA from *Bubalus bubalis* not conserved evolutionarily is transcribed. *Z Naturforsch C* 59(11-12):874-879.

[10] Bishop, R., Morzaria, S., & Gobright, E. (1998). Linkage Of Two Distinct At-Rich Minisatellites At Multiple Loci In The Genome Of *Theileria Parva*. *Gene* 216, 245-254

[11] Blott, S.C., Williams, J.L., & Haley, C.S. (1999). Discriminating among cattle breeds using genetic markers. *Heredity (Edinb)* Pt 6:613-619.

[12] Boeva,V., Regnier, M., Papatsenko, D., Makeev, V. (2006). Short fuzzy tandem repeats in genomic sequences, identification, and possible role in regulation of gene expression. *Bioinformatics* 2006 Mar 15; 22(6):676-684.

[13] Bromham, L. (2002).The Human Zoo: Endogenous Retroviruses in the Human Genome. *Trends Ecol. Evol.* 17, 91–97.

[14] Brown, T. A. (2002). The Repetitive DNA Content Of Genomes. *Genomes* 59-64.

[15] Capy P. (1998). Evolutionary biology. A plastic genome. Nature 396(6711):522-3.

[16] Catasti, P., Chen, X., Mariappan, S.V., Bradbury, E.M. & Gupta, G. (1999) DNA Repeats In The Human Genome. Genetica 106, 15-36.

[17] Cavalier-Smith, T. 1985. Eukaryotic Gene Numbers, Non-Coding DNA, And Genome Size. *In* The Evolution Of Genome Size. *Edited By* T. Cavalier-Smith. *John Wiley And Sons, Chichester, U.K.* Pp. 69–103.

[18] Charlesworth, B., Sniegowski, P., & Stephan, W. (1994). The Evolutionary Dynamics Of Repetitive DNA In Eukaryotes. *Nature* 371, 215-220.

[19] Chattopadhyay, M., Prashant, S. G., Kapur, V., Azfer, Md. A., Prakash, B., & Ali, S. (2001). Satellite Tagged Transcribing Sequences In The Bubaline *Bubalus bubalis* Genome Undergo Programmed Modulation In The Meiocytes: Possible Implication In Transcriptional Inactivation. *DNA Cell Biol.* 20, 587-593.

[20] Debrauwere, H., Gendrel, G.C., Lechat, S., & Dutreix, M. (1997). Differences and similarities between various tandem repeat sequences: Minisatellites and microsatellites. *Biochimie.* 79:577–586.

[21] Dey, I., & Rath, P.C. (2005). A Novel Rat Genomic Simple Repeat DNA With RNA-Homology Shows Triplex (H-DNA)-Like Structure And Tissue-Specific RNA Expression. *Biochem. Biophys. Res. Commun.* 276, 286-228.

[22] Durward, E., Shiu, O.Y., Luczak, B., & Mitchelson, K. R. (1995). Identification Of Clones Carrying Minisatellite-Like Loci In An Arabidopsis Thaliana Yac Library. *Journal Of Experimental Botany* 46, 271-274.

[23] Elgar, G., & Vavouri, T. (2008). Tuning in to the signals: noncoding sequence conservation in vertebrate genomes. *Trends Genet.* **24** (7): 344–52.

[24] Epplen, J.T., Kyas, A. & Maueler, W. (1996) Genomic Simple Repetitive DNAs Are Targets For Differential Binding Of Nuclear Proteins. *Febs Letters* 389, 92-95.

[25] Feng, Y., Yang, W., Ryan, U., Zhang, L., Kvác, M., Koudela, B., Modry, D., Li, N., Fayer, R., & Xiao, L. (2011). Development of a multilocus sequence tool for typing Cryptosporidium muris and Cryptosporidium andersoni. *J Clin Microbiol.* 49(1):34-41.

[26] Fondon, J.W., & Garner, H.R. (2004) Molecular Origins Of Rapid And Continuous Morphological Evolution. *Proc. Natl. Acad. Sci. USA* 101, 18058–18063.

[27] Furano, A. V. (2000). The Biological Properties And Evolutionary Dynamics Of Mammalian Line-1 Retrotransposons Prog. *Nucleic Acid Res.* 64,255-294.

[28] Georges, M., & Andersson, L. (1996). Livestock Genomics Comes Of Age. Genome Res. 6,907-921.

[29] Georges, M., Gunawardana, A., Threadgill, D.W., & Lathrop, M. (1991). Characterization Of A Set Of Variable Number Of Tandem Repeat Markers Conserved In Bovidea. *Genomics* 11, 24–32.

[30] Gur-Arie, R., Cohen, C.J., Eitan, Y., Shelef, L., Hallerman, E.M. & Kashi, Y. (2000) Simple Sequence Repeats In Escherichia Coli: Abundance, Distribution, Composition, And Polymorphism. *Genome Res.* 10, 62–71.

[31] Haber, J.E., & Louis, E.J. (1998). Minisatellite Origins In Yeast And Humans. *Genomics* 48, 132-135.

[32] Han, J.S., Boeke, J.D. (2005). LINE-1 retrotransposons: modulators of quantity and quality of mammalian gene expression? *Bioessays* **27**:775-784.

[33] Han, J.S., Szak, S.T., & Boeke, J.D. (2004). Transcriptional disruption by the L1 retrotransposon and implications for mammalian transcriptomes. *Nature* 429:268-274.

[34] Harris, A.S., & Wright, J.M. (1995). Nucleotide sequence and genomic organization of cichlid fish minisatellites. *Genome* 38(1):177-184.

[35] Henikoff, S., Ahmad, K., & Malik, H. S. (2001). The Centromere Paradox: Stable Inheritance With Rapidly Evolving DNA. *Science* 293, 1098–1102.

[36] Hochgeschwender, U., & Brennan, M.B. (1991) Identifying Genes Within The Genome: New Ways For Finding The Needle In A Haystack. *Bioessays* 13, 139-144.

[37] International Human Genome Sequencing Consortium. (2004) Finishing the euchromatic sequence of the human genome. Nature 431(7011):931-945.

[38] Jarman, A.P., & Wells, R.A. (1989). Hypervariable Minisatellites: Recombinators Or Innocent Bystanders? *Trends Genet.* 5,367–371.

[39] Jeffreys, A. J., Neil, D. L., & Neumann, R. (1998). Repeat Instability At Human Minisatellites Arising From Meiotic Recombination. *Embo J.* 17, 4147–4157.

[40] Jeffreys, A.J., & Wilson, V., & Thein, S.L. (1985). Hyper Variable 'Minisatellite' Regions In Human DNA. *Nature* 314, 67-73.

[41] Jobling, M.A., & Tyler-Smith, C. (2003). The human Y chromosome: an evolutionary marker comes of age. *Nat Rev Genet.* 4(8):598-612.

[42] Jordan, I. K., Rogozin, I. B., Glazko, G.V. & Koonin, E.V.(2003). Origin of a substantial fraction of human regulatory sequences from transposable elements. Trends in Genetics 19, 68–72.

[43] Kapur, K., Prasanth, S. G., O'ryan, C., Azfer, Md, A., & Ali, S. (2003). Development Of A DNA Marker By Minisatellite Associated Sequence Amplification (Masa) From The Endangered Indian Rhino (*Rhinoceros Unicornis*). *Mol. Cell. Probes* 17, 1-4.

[44] Kelly, M.K., Alver, B., & Kirkpatrick, D.T. (2011). Minisatellite alterations in ZRT1 mutants occur via RAD52-dependent and RAD52-independent mechanisms in quiescent stationary phase yeast cells. DNA Repair (Amst). 10(6):556-566.

[45] Kim, M. & Mullet, J.E. (1995) Identification Of A Sequence-Specific DNA Binding Factor Required For Transcription Of The Barley Chloroplast Blue Light-Responsive Psbd-Psbc Promoter. *Plant Cell* 7, 1445-1457.

[46] Kit, S. (1961). Equilibrium Sedimentation In Density Gradients Of DNA Preparations From Animal Tissues. *J. Mol. Biol.* 3,711-716.

[47] Kumar, S., Gupta, R., Kumar, S., & Ali, S. (2011). Molecular mining of alleles in water buffalo *Bubalus bubalis* and characterization of the TSPY1 and COL6A1 genes. *PLoS One* 6(9):e24958.

[48] Lander, E.S., Linton, L.M., Birren, B., Nusbaum, C., Zody, M.C., Baldwin, J., Devon, K., Dewar, K., Doyle, M., Fitzhugh, W., Funke, R., Gage, D., et al. (2001). Initial Sequencing and Analysis of the Human Genome. *Nature* 409, 860–921.

[49] Legendre, M., Pochet, N., Pak, T., & , K.J. (2007). Sequence-based estimation of minisatellite and microsatellite repeat variability. *Genome Res.* 17(12):1787-1796.

[50] Lopes, J., Ribeyre, C., & Nicolas, A. (2006). Complex minisatellite rearrangements generated in the total or partial absence of Rad27/hFEN1 activity occur in a single generation and are Rad51 and Rad52 dependent. *Mol Cell Biol.* 26(17):6675-6689.

[51] Lue, N. F., Buchman, A. R., & Kornberg, R. D. (1989) Activation Of Yeast RNA Polymerase Ii Transcription By A Thymidine-Rich Upstream Element In Vitro. *Proc. Natl. Acad. Sci. USA* 86, 486-490.

[52] Martienssen, R. A., & Colot, V. (2001). DNA Methylation And Epigenetic Inheritance In Plants And Filamentous Fungi. *Science* 293, 1070-1074.

[53] Miller, W. J., & Capy, P., eds. (2004), Mobile genetic elements: protocols and genomic applications, *Humana Press*, 289.

[54] Mudgal, V.O. (1988). Comparative Efficiency For Milk Production Of Buffaloes And Cattle In The Tropics. *Proceedings Of Ii World Buffalo Congress.* New Delhi, India, Vol Ii, Part Ii, 454–462.

[55] Nakamura, Y., Leppert, M., O'connell, P., Wolff, R., Holm, T., Culver, M., Martin, C., Fujimoto, E., Hoff, M., Kumlin, E. et al. (1987).Variable Number Of Tandem Repeat (Vntr) Markers For Human Gene Mapping. *Science* 235, 1616–1622.

[56] Norman, A., D. (2001). APPENDIX 1B Overview of Human Repetitive DNA Sequences. Current Protocols in Human Genetics.

[57] Ohno, S. (1972). So Much `Junk' In Our Genomes. *Brookhaven Symp. Biol.* 23, 366-370.

[58] Pathak, D., & Ali, S. (2011). RsaI repetitive DNA in Buffalo *Bubalus bubalis* representing retrotransposons, conserved in bovids, are part of the functional genes. *BMC Genomics* 12:338.

[59] Pathak, D., Srivastava, J., Premi, S., Tiwari, M., Garg, L.C., Kumar, S., & Ali, S. (2006). Chromosomal localization, copy number assessment, and transcriptional status of *BamHI* repeat fractions in water buffalo Bubalus bubalis. *DNA Cell Biol.* 25(4):206-214.

[60] Pathak, D., Srivastava, J., Samad, R., Parwez, I., Kumar, S., & Ali, S. (2010). Genome-wide search of the genes tagged with the consensus of 33.6 repeat loci in buffalo *Bubalus bubalis* employing minisatellite-associated sequence amplification. *Chromosome Res.* 18(4):441-458.

[61] Patience, C., Takeuchi, Y., & Weiss, R. A. (1997). Infection Of Human Cells By An Endogenous Retrovirus Of Pigs. *Nat. Med.* 3, 276-282.

[62] Proudfoot, N.J., Gill, A., & Maniatis, T. (1982). The Structure Of The Human Zeta-Globin Gene And A Closely Linked, Nearly Identical Pseudogene. *Cell* 31, 553-563.

[63] Rawal, L., Ali, S., & Ali, S. (2012). Molecular mining of GGAA tagged transcripts and their expression in water buffalo Bubalus bubalis. *Gene* 492(1):290-295.

[64] Reed, J.M., Fleischer, R. C., Eberhard, J., & Oring, L.W. (1996). Minisatellite DNA Variability In Two Populations Of Spotted Sandpipers Actitis Macularia In Minnesota, U.S.A. Wader Study Group Bull. 79, 115-117.

[65] Richard, C. & Mark, B. (2009). The impact of retrotransposons on human genome evolution. *Nature Reviews Genetics* 10 (10): 691–703.

[66] Roy-Engel A, M., Carroll, M.L., Vogel, E., et al. (2001). Alu insertion polymorphisms for the study of human genomic diversity. *Genetics* 159 (1): 279–290.

[67] Royle, J.R., Clarkson, R.E., Wong, Z., & Jeffreys, A.J. (1988). Clustering Of Hypervariable Minisatellites In The Proterminal Regions Of Human Autosomes. *Genomics* 3, 352–360.

[68] Sethi, R. K. (2003). Improving Rivering And Swamp Buffaloes Through Breeding. 4th Asian Buffalo Congress Lead Papers, 50.

[69] Shapiro, J.A., & Von Sternberg, R.(2005). Why repetitive DNA is essential to genome function. *Biol Rev Camb Philos Soc.*80(2):227-50

[70] Sinden, R.R. (1999). Biological Implications Of The DNA Structures Associated With Disease-Causing Triplet Repeats. *Am. J. Hum. Genet.* 64, 346–353.

[71] Skuce, R. A., Mccorry, T. P., Mccarroll, J. F., Roring, S.M.M., Scott, A.N., Brittain, D., Hughes, S.L., Hewinson, R.G., Sydney, D., & Neil, L. (2002). Discrimination Of Mycobacterium Tuberculosis Complex Bacteria Using Novel Vntr-Pcr Targets. *Microbiology* 148, 519-528.

[72] Slamovits, C.H., & Rossi, M.S. (2002). Satellite DNA: Agent Of Chromosomal Evolution In Mammals. Mastozoología Neotropical 9, 297-308.

[73] Srivastava, J., Premi, S., Kumar, S., & Ali, S. (2008). Organization and differential expression of the GACA/GATA tagged somatic and spermatozoal transcriptomes in Buffalo Bubalus bubalis. *BMC Genomics* 9:132.

[74] Srivastava, J., Premi, S., Pathak, D., Ahsan, Z., Tiwari, M., Garg, L.C., & Ali, S. (2006). Transcriptional status of known and novel genes tagged with consensus of 33.15 repeat loci employing minisatellite-associated sequence amplification (MASA) and real-time PCR in water buffalo, Bubalus bubalis. *DNA Cell Biol.* 25 (1):31-48.

[75] Stoker, N.G., Cheah, K.S.E., Griffin, J.R., Pope, F.M., & Solomon, E. (1985). A Highly Polymorphic Region 3' To The Human Type Ii Collagen Gene. *Nucleic Acids Res.* 13, 4613–4622.

[76] Subramanian, V.M., Madgula, R., George, R.K., Mishra, M.W., Pandit, C.S. Kumar, & L. Singh. (2003). *Bioinformatics* 19, 549–552.

[77] Sykorová, E., Fajkus, J., Mezníková, M., Lim, K.Y., Neplechová, K., Blattner, F.R., Chase, M.W., & Leitch, A.R. (2006). Minisatellite telomeres occur in the family Alliaceae but are lost in Allium. *Am J Bot.* 93(6):814-823.

[78] Tang, S.J. (2011). Chromatin Organization by Repetitive Elements (CORE): A Genomic Principle for the Higher-Order Structure of Chromosomes. *Genes* 2011, 2(3), 502-515

[79] Tautz, D. (1989) Hypervariability of simple sequences as a general source for polymorphic DNA markers. *Nucleic Acids Res.* 17(16):6463-6471.

[80] Tautz, D. (1993). Notes On The Definition And Nomenclature Of Tandemly Repetitive DNA Sequences. In DNA Fingerprinting: State Of The Science. 21–28.

[81] Tautz, D., & Renz, M. (1984). Simple Sequences Are Ubiquitous Repetitive Components Of Eukaryotic Genomes. Nucleic Acids Res. 25, 4127-4138.

[82] Tomilin, N.V. (2008). Regulation Of Mammalian Gene Expression By Retroelements And Non-Coding Tandem Repeats. *Bioessays* 30, 338–348.

[83] Toth, G., Gaspari, Z. & Jurka, J. (2000). Microsatellites In Different Eukaryotic Genomes: Survey And Analysis. *Genome Res.* 10, 967–981.

[84] Tourmente, S., Deragon, J.M., Lafleuriel, J., Tutois, S., Pelissier, T., Cuvillier, C., Espagnol, M.C., & Picard, G. (1994). Characterization Of Minisatellites In Arabidopsis Thaliana With Sequence Similarity To The Human Minisatellite Core Sequence. Nucleic Acids Research 22, 3317–3321.

[85] Ugarkovic, D., & Plohl, M. (2002). Variation In Satellite DNA Profiles-Causes And Effects. *EMBO* 21, 5955-5959.

[86] Vermaak, D., Bayes, J.J., & Malik, H.S. (2009). A surrogate approach to study the evolution of noncoding DNA elements that organize eukaryotic genomes. *J Hered.* 100(5):624-636.

[87] Vinces, M.D., Legendre, M., Caldara, M., Hagihara, M., & , K.J. (2009). Unstable tandem repeats in promoters confer transcriptional evolvability. *Science* 324(5931):1213-1216.

[88] Walters, R. (2011). More commercial benefits on horizon as pig genome project nears completion. Information from pig genome already being used in the industry. *Pig International* , 41, 2:16

[89] Wilson, G.A., & Strobeck, C. (1999). The isolation and characterization of microsatellite loci in bison, and their usefulness in other artiodactyls. *Anim Genet.* 30(3):226-227.

[90] Zhang, L., Yuan, D., Yu, S., Li, Z., Cao, Y., Miao, Z., Qian, H., & Tang, K. (2004). Preference of simple sequence repeats in coding and non-coding regions of Arabidopsis thaliana. *Bioinformatics* 20(7):1081-1086.

Chapter 8

Medicago truncatula Functional Genomics – An Invaluable Resource for Studies on Agriculture Sustainability

Francesco Panara, Ornella Calderini and Andrea Porceddu

Additional information is available at the end of the chapter

1. Introduction

Legume functional genomics has moved many steps forward in the last two decades thanks to the improvement of genomics technologies and to the efforts of the research community. Tools for functional genomics studies are now available in *Lotus japonicus*, *Medicago truncatula* and soybean. In this chapter we focus on *M.truncatula*, as a model species for forage legumes, on the main achievements obtained due to the reported resources and on the future perspectives for the study of gene function in this species.

2. Why do we need a functional genomics tool for forage legumes?

Legumes are widely grown for grain and forage production, their world economic importance being second only to grasses. Legume species are unique among cultivated plants for their ability to carry out endosymbiotic nitrogen fixation with rhizobial bacteria, a process that takes place in a specialized structure known as nodule. Moreover legumes are able to establish other types of interactions such as arbuscular mycorrhyzal symbiosis with several fungi. For these outstanding biological properties legumes are considered among the most promising species for improving the sustainability of agricultural systems. In fact for farming system to remain productive and to be environmentally and economically sustainable on the long term it is necessary to replenish the reserves of nutrients which are removed or lost from the soil. Nodulating legumes have the potential to provide all nitrogen required for their growth and in this way to influence its balance in the soil and thus its availability for subsequent crops. In addition by reducing the inputs of fertilizers, legumes reduce the risk of nitrogen contamination of water resources. Furthermore, probably due to the wealth of interactions with other organisms, legumes have evolved an intricate network of secondary metabolites that, as the recent advances in the knowledge of their nutraceutical properties are proving, can be considered of great importance for livestock welfare and for the quality of their products.

Two legume species, *Medicago truncatula* and *Lotus japonicus*, are being used as model to study legume genetics and genomics. *Medicago truncatula* is closely related to alfalfa, the most important forage legume in the world. It has a small, diploid genome, it is self fertile and amenable to genetic transformation. In the present review we summarize the state of the art of *M. truncatula* genomics with particular emphasis on the available resources for functional genomics studies such as mutant collections.

3. *Medicago truncatula* genome sequencing

Functional genomics is greatly aided by knowledge on genome sequence and transcriptome of the target species. A concerted effort was carried out in *M. truncatula* which made genome data available to the community.

Legumes are the plant family with the greatest amount of genomic data available. Three legume species, *Medicago truncatula*, *Lotus japonicus*, and *Glycine max*, have been sequenced (1). The assembly of *Medicago truncatula* genome is close to completion (2).

M. truncatula sequencing was initially carried out on Bacterial Artificial Chromosome (BAC) libraries following a BAC-by-BAC approach focused on gene-rich BACs. To date the available sequence data consist of three main batches: i) 246Mb of non redundant sequences that could be organized in large scaffolds separated by gaps and anchored to the eight *M.truncatula* physical chromosomes, ii) 17.3Mb of unanchored scaffolds and iii) 104.2Mb of additional unique sequence obtained by next generation sequencing (NGS) with Illumina sequencing. In total, 367.5Mb of *M.truncatula* genome representing 73,5% of the ~500Mb of the predicted genome size and about 94% of the expressed genes is available.

Taken together BAC sequences and non-redundant Illumina assemblies contain 62,388 gene loci with 14,322 gene prediction annotated as transposons. The average *M. truncatula* gene is 2,211 bp in length, contains 4.0 exons and has a coding sequence of 1,001 bp. Genome analysis and comparisons to other sequenced genomes allowed the identification of a 58-Myr-ago whole genome duplication (WGD) that has been associated with the evolution of rhizobial nodulation in *M. truncatula* and its relatives. Some nodulation-specific signalling components might have evolved through duplication and neo-functionalization from more ancient genes involved in host-mycorrhyzal signalling (2).

Another interesting feature is the presence of many amplified and somehow specialized gene families like nine leghaemoglobines, 563 Nodule Cystein-Rich Peptides (NCRs), 764 nucleotide-binding site and leucine-rich repeat (NBS-LRR) genes, genes in the flavonoid pathway such as chalcone synthases (CHS), chalcone reductases, chalcone isomerases. Many gene duplications occurred with the creation of large gene clusters (2).

The availability of a first draft of the *Medicago truncatula* genome sequence has promoted several initiatives aimed at identifying molecular markers suitable for both evolutionary and genetic mapping studies.

384 inbred lines of *M.truncatula* with a 5x coverage and a subset of 30 with deep coverage (20x) will be resequenced through an Illumina-Solexa sequencing pipeline by the *Medicago* HapMap Project. In the first report on the analysis of sequence data from 26 *M.truncatula* accessions with ~15x average genome coverage, 3,063,923 mapped single nucleotide polymorphisms (SNPs) were described and first estimates of nucleotide diversity (θw =0.0063 and $\theta\pi$ =0.0043 bp–1), population scaled recombination rate and rate of decay of linkage disequilibrium have been published (3). More recently the same material was used to estimate population recombination rates at 1 kb scale and very interestingly in the three chromosomes analysed recombination was higher near centromeric regions in stark contrast to what observed in every non-plant system and in the majority of plants that show a negative gradient of recombination from telomeric to centromeric regions (4).

In parallel, plant phenotyping is ongoing in greenhouse experiments for the *Medicago* lines. The combination of genetic and phenotypic data will be organized in a platform for genome-wide association mapping (GWAS) studies.

4. Functional genomics in *Medicago truncatula*

The discovery of gene function in model species is accomplished by exploiting resources such as mutant collections, using the ability to implement plant genetic transformation and analyzing gene transcription. In *M. truncatula* all the three approaches can be performed.

Several strategies were pursued in *M. truncatula* to produce mutant collections (Tab. 1) and they will be analysed in the following section.

Reference	Mutagenesis technique	Background	Notes
(5)	EMS	Jemalong population 2828	1,500 seeds treated for 24h with 0.2%EMS. 400 M1 plants obtained. 250,000 M2 seeds harvested as a single batch.
(6)	γ-ray	Jemalong line J5	462 M1 plants, screened as M2 families.
(7)	Ethyl-nitrosourea		
(8)	EMS	Jemalong A17	3-7,000 M1 plants in 10-20 lots. M2 seeds bulked from each lot.
(9)	T-DNA	R108-1 (c3)	Test populations with 3 different T-DNAs
(10)	Tnt1	R108-1 (c3)	First test population with Tnt1 (~200 R0 plants)
(11)	FNB	Jemalong A17	80.000 M1, 460.000 M2 http://bioinfo4.noble.org/mutant/
(12;13)	Tnt1	R108-1 (c3)	7,000 Tnt1 mutants, presently extended to 19,000 as reported in http://bioinfo4.noble.org/mutant/

Reference	Mutagenesis technique	Background	Notes
(14)	Tnt1	R108-1 (c3)	~1000 R0
(14)	EMS	Jemalong 2HA	2500 M2 plants
(14)	Activation tagging	R108-1 (c3)	~100 mutant lines
(15)	EMS	Medicago littoralis 'Angel'	Development of new annual medics varieties (i.e. resistant to herbicides)
(16)	Tnt1	Jemalong 2HA	Mutants produced in the frame of the european Grain Legumes Integrated Project (GLIP). The total number of mutants produced by 10 labs all around Europe should be several thousands (~6000). 2000 of them will integrate the Tnt1 collection at Noble.
(17)	FNB	Jemalong A17	31,200 M1 plants, 156,000 M2 plants
(18)	EMS	Jemalong A17	http://195.220.91.17/legumbase/ 2 populations. The first (not using single seed descent, SSD) 500 M1 produced 4500 M2. In the second (using SSD) 4350 M1 and 4350 M2.

Table 1. Mutant collections in *Medicago* spp. **EMS = ethyl methanesulphonate.**

5. Chemical-physical mutagenesis

5.1. Target Induced Local Lesion IN Genomes (TILLING)

Alkylating agents such as ethyl methanesulphonate (EMS) have been used to develop mutant collections of *Medicago truncatula*. EMS induces single base pair C/G to A/T substitution in nucleotides. The mutagenized seeds are germinated and the resulting plants are selfed to produce M1 progenies. The M1 plants are then grown and a TILLING M2 collection is established by growing few seeds from each M1 plant. Total genomic DNA is purified from each M1 plant and pooled. The mutant collections are usually screened with reverse genetics approaches. TILLING involves the identification of mismatches in heteroduplexes formed by single stranded DNA from the wild type and mutant alleles of the target locus. The target sequences are generated by PCR amplification from bulked DNA isolated from single M1 plants using labelled primers appropriate for the detection strategy employed. The amplicons are then heated, causing strand separation, re-annealed in order to form heteroduplexes, cleaved by an endonuclease active on single stranded DNA (i.e. CelI from celery) at the mismatch point and the products separated by electrophoresis. Several EMS mutant collections of *Medicago truncatula* are available. Within the framework of the European Grain Legume Integrated Project two mutant collections were established. The two collections showed the same number of M2 lines that however were obtained from

M1 populations with different size. Genetic analysis of the two collections allowed to define that the number of meristematic cells that contribute to seed (germ-line) in *Medicago truncatula* is 3. The number of mutations detected in the two EMS populations was 1 every 485 kb. A pilot reverse genetic experiment with 56 target genes revealed an efficiency of 13 independent alleles per exon screened, 67% of which were missense and 5% nonsense mutations. An Italian functional genomics initiative produced a small collection of TILLING mutants with about 2500 M2 lines and a reported efficiency of about 4 independent alleles for target sequence. Catalogue of mutant phenotypes were developed and services for reverse screening with target sequence are available (http://inra.fr/legumbase). A list of *M.truncatula* mutants is reported in Table 2.

Reference	Mutagenesis technique	New mutants	Phenotype	Gene/Mutant line
(5)	EMS	1	Nod+Fix-	TE7=Mt*sym*1
(6)	γ-ray	2	Nod-, Myc-	TR25, TR26
(6)	γ-ray	4	Nod±, Myc+	TR34, TR79, TR89, TRV9
(6)	γ-ray	9	Nod+Fix-, Myc+	TR3, TR9, TR13, TR36, TR62, TR69, TR74, TR183, TRV15
(6)	γ-ray	3	Nod++Nts, Myc+	TR122, TRV3, TRV8
(8)	EMS	1	EIN, Nod++	*sickle* = *skl1*
(8)	EMS	1	Nod-	C71 = *Domi*
(19)	γ-ray	1	Nod-	TRV25
(20)	EMS	3	developmental	*mtapetala (tap), palmyra (plm), speckle (spk)*
(21)	EMS	5	Nod-	B85, B129, C54, P1, Y6 Individuation of 4 complementation groups (DMI1, DMI2, DMI3, NSP): *dmi1-1* = C71 = *domi* *dmi1-2* = B129 *dmi1-3* = Y6 *dmi2-1* = TR25 *dmi2-2* = TR26 *dmi2-3* = P1 *dmi3-1* = TRV25 *nsp1-1* = B85 *nsp1-2* = C54
(22)	EMS	7	Calcium oxalate defective	*cod1, cod2, cod3, cod4, cod5, cod6, cod7*
(23)	EMS	1	Nod-, root hair deformation	*hcl* = *B56*
(24)	EMS	1	Blocked in the formation of nodule primordial	*pdl1*

Reference	Mutagenesis technique	New mutants	Phenotype	Gene/Mutant line
(24)	EMS	1	Blocked in the formation and/or maintenance of epidermal cell infection	*lin1* (first citation)
(21)	EMS	3	Nod-, root hair deformations	*hcl-1* = B56 *hcl-2* = W1 *hcl-3* = AF3
(25)	EMS	6	Oxalate crystal morphology defective	*cmd1, cmd2, cmd3, cmd4, cmd5, cmd6*
(26)	Fast Neutron	2	Nod-, cortical cell division	*nsp2-1* *nsp2-2*
(27)	EMS	1	Nod-, does not respond to Nod Factors by induction of root hair deformation	*nfp* = C31
(28)	EMS	1	Nod++	*Sunn*
(29)	EMS	1	Blocked in the formation and/or maintenance of epidermal cell infection	*lin1* (first description)
(30)	EMS	1	Numerous infections and polyphenolics	*Nip*
(31)	EMS	1	Nod-, defective in lateral root development	*Latd*
(32)	γ-ray	2	Nod+,Fix-,Myc+	*Mtsym20* = TRV43, TRV54 *Mtsym21* = TRV49
(32)	γ-ray	1	Nod-/+,Myc+	*Mtsym15* = TRV48;
(32)	γ-ray	1	Nod-,Myc-/+	*Mtsym16* = TRV58
(33)	Fast Neutron	6	Fix-	*dnf1-1* = 1D-1; *dnf1-2* = 4A-17; *dnf2* = 1B-5; *dnf3* = 2C-2; *dnf4* = 2E-1; *dnf5* = 2F-16; *dnf6* = 2H-8; *dnf7* = 4D-5
(34)	Fast Neutron	1	Nod-	*bit1*
(35)	Tnt1	1	Single leaflet	*sgl1*
(36)	Fast Neutron	1	Increased nodule number	*Efd*
(37)	EMS	1	Impaired in nodule primordium invasion	*Api*
(38)	EMS	1	Aberrant root hair curling and infection thread formation	*Rpg*
(39)	EMS	1	Myc++, Nod-/+	B9

Reference	Mutagenesis technique	New mutants	Phenotype	Gene/Mutant line
(40)	Fast neutron	1	Leaf dissection	*palm1*
(41)	T-DNA	1	Compact roots	*cra1* (not tagged)
(42)	Tnt1	various	Leaf epidermal morphology	Various
(43)	Tnt1	1	Lack of lignin in the interfascicular region	*nst1*
(44)	Tnt1	1	Secondary cell wall thickening in pith	*mtstp1*
(45)	Fast neutron	1	Compund leaf development	*fcl1*
(46)	Fast neutron	1	Root determined noudulation	*rdn1*
(47;48)	Fast neutron	1	Myc-, Nod-	*Vpy*
(49)	Activation tagging	1	Lack of hemolytic saponins	*Lha*
(50)(1)	Tnt1	1	Stay green	*MtSGR*
(51)	Tnt1	1	Smooth leaf margin	*slm1*
(52)	Tnt1	1	Reduced leaf blade expansion	*Stf*
(53)	Tnt1	1	Inhibition of rust germ tube differentiation	*irg1*

Table 2. *Medicago truncatula* mutants. Nodulation phenotypes: Nod++ = hypernodulator, Nod+ = wild type nodulator, Nod± = reduced nodulation, Nod- = lack of nodules, Nod-/+ = late nodulation. Nitrogen fixation phenotypes: Fix+ = wild type, Fix- = no fixation. Mychorrhization phenotypes: Myc+ = wild type, Myc- = absent or reduced mychorrhizas, Myc-/+ = mix of normal colonization and events of formation only of appressoria with no intercellular hyphae developing from them, Myc++ = hyper responsive to mychorrhization. Nts = nitrate tolerant nodulation. EIN = ethylene insensitive.

5.2. "Delete a gene" collections

Irradiation of plant seeds to appropriate dose of fast neutrons and γ-rays results in deletion of DNA fragments of variable lengths with an average modest reduction of seed viability.

Large mutant collections by seed irradiation have been created for *Medicago truncatula* functional genomics studies. Although the first experiments were based on γ-ray irradiation of the Jemalong J5 seeds (6) the main body of the collection was obtained by Fast Neutron Bombardment (FNB) of the genotype A17. Globally the two larger collections, stored at the John Innes Center and at the Noble Foundation, consist of about 616,000 M2 FNB families.

Both reverse and forward genetic approaches have been successfully applied to study mutants from these collections.

Reverse screening of FNB populations have been carried out by the DeTILLING strategy described by Rogers C *et al.* (17). This strategy allows detection of mutants by PCR on bulks of DNAs of FNB mutants. The wild type target amplification is avoided by a strategy that combines restriction enzyme digestion of the template and the use of poison primers. With this strategy a mutant recovery rate of 29% has been obtained from a population of 156.000 M2 plants (4 genes out of 14 screened).

Nevertheless deletion size can hamper reverse genetics screening (Chen R., personal communication) leaving forward genetics as the main choice in case of FNB populations. However map-based cloning required to discover the mutation of interest is helped by strategies such as transcriptional cloning, originally devised by Mitra R M *et al.* (54), which has allowed the identification of FNB induced mutations (see Table 3). This approach relies on the identification of mutated genes through detailed genome-wide transcriptomic analyses. Also genome-wide analyses of FNB mutant are expected to benefit of the recent development of a *Medicago truncatula* genome-wide tiling array by Nimblegen. A list of *Medicago truncatula* FNB mutants characterized by forward genetics approaches is reported in Table 2.

6. Insertional mutagenesis with DNA mobile elements

6.1. Tnt1

T-DNA tagging has been the strategy of choice for many mutant collections in *Arabidopsis* and it has allowed fundamental discoveries in gene functions and advances in both basic and applied plant research (55). Unfortunately only *Arabidopsis* can be transformed easily by the floral-dip method which allows the generation of large numbers of mutants in a cost-effective manner. Up to now transformation for the other plant species including *M. truncatula* can only be achieved by tissue culture-based protocols requiring great efforts to produce the number of mutants that would allow a significant genome coverage. An interesting strategy has been recently published in the legume *Lotus japonicus* based on the endogenous retrotransposon LORE1 (56;57). LORE1, originally activated via tissue culture, retained its activity for some regenerated plants in the subsequent generations. Based on such discovered germline activity, tagged M1 mutant collections were produced by seed propagation from activated starter lines (M0) (57;58).

In *M.truncatula* large scale collections of mutants have been constructed using the tobacco Tnt1 retrotransposon. d'Erfurth and colleagues have demonstrated that in the *Medicago truncatula* R108 genotype, this element has the ability to transpose during the early steps of in vitro regeneration (10) with a high rate of insertion in transcribed genomic regions. Sequence analyses of insertion sites has showed the virtual absence of insertion site preference. The average amount of new insertions per regenerated line was calculated in the order of ~25. Based on these data it was shown that a collection of 14-16.000 Tnt1 lines will store tagging events for about 90% of *M.truncatula* genes (13). Such an ambitious objective has been pursued by working on two *Medicago truncatula* lines.

The collection maintained at the Noble Foundation (http://bioinfo4.noble.org/mutant/) which includes also the first mutants generated by CNRS in France, is based on the genotype R108-c3. Another collection of about 1000 lines from the same R-108 line was produced by CNR-IGV in Italy.

In the framework of the GLIP project 8000 Tnt1 mutants were produced from the Jemalong 2HA (2HA3-9-10-3) line. The GLIP collection is maintained by the various labs that participated to the project and a subset of plants were merged with the collection at the Noble Foundation.

Iantcheva and colleagues reported that Tnt1 transposition efficiency in Jemalong 2HA has a lower efficiency with only 10-15 new insertions per line and a variable percentage of regenerated plants without transposition (16). The adoption of 2HA line for mutagenesis instead of R108, was motivated by the highest DNA homology to the line used for genome sequencing (Jemalong A17), and for the presence of active and characterized endogenous retroelements (59).

Tnt1 mutant collections have been screened with both forward and reverse genetic approaches. Forward approaches have been based on cloning of host sequence flanking the insertion sites and subsequent identification of events linked to the studied mutation. Based on the duplicated Tnt1 long terminal repeats (LTR) sequences several molecular approaches including thermal asymmetric interlaced (TAIL)-PCR, Inverse-PCR have been used to recover the host sequences flanking the insertion sites (60). Segregation analysis of each cloned insertion site can then be used to select the event linked to the mutation. In alternative the insertion sites associated with the mutations can be selected by segregation analysis prior to host sequence cloning by employing a sequence specific amplification polymorphism (S-SAP) based protocol.

Confirmation of the identity of the mutation can be obtained by means of complementation tests based on the reintroduction of the wild type gene sequence in the mutated background. In alternative one could obtain independent alleles of the target gene and compare their similarity to the original mutant phenotype. This can be done using TILLING and Tnt1 mutant populations as demonstrated by many publications that report successful recovery of alleles by reverse screening (61) and Table 3. The power of the Tnt1 mutagenesis approach is also witnessed by the prevalence of publications reporting successful gene cloning based on such strategy compared to the others since 2008 (Table 2 and 3).

Reference	Mutant	Gene	Approach
(62)	dmi2	NORK	Physical mapping
(63)	dmi1	AY497771, possible membrane receptor	Physical mapping
(64)	dmi3	Ca_2 and Calmodulin dependent protein kinase	Physical mapping/Transcriptional based cloning

Reference	Mutant	Gene	Approach
(65)	nsp2	GRAS Transcriptional regulator	Physical mapping
(66)	nsp1	GRAS Transcriptional regulator	Physical mapping
(46)	sunn	CLV1-like LRR receptor kinase	Physical mapping and gene homology
(67)	mtpim	MADS-box	Reverse screening on Tnt1 collection
(68)	mtpt4	Phosphate transporter	RNAi and TILLING (reverse)
(34)	bit1	ERF transcription factor required for nodulation (ERN)	Transcriptional based cloning
(35)	sgl1	MtUNI (transcription factor)	Tnt1 forward
(36)	efd	Ethilene responsive factor required for nodule differentiation	Fast neutron reverse
(38)	rpg	Putative long coiled-coil protein	Map based cloning
(28)	sickle	MtEIN2, ethylene signaling gene	Map based cloning and gene homology
(69)	lin	E3 ubiquitin ligase containing a U-box and WD40 repeat domains	Positional cloning
(70)	srlk	LRR kinase	TILLING reverse and RNAi
(71)	mate1	MATE	Tnt1 reverse
(72)	ugt78g1	Glucosyl transferase	Tnt1 reverse
(73)	mtapetala	MtPI, MADS Box transcription factor	RNAi and mutation segregation analisys
(40)	palm1	Cys(2)His(2)zinc finger transcription factor	Fast neutron forward and Tnt1 reverse
(74)		MtSYMREM1, remorin	Tnt1 reverse
(44)	dnf1	Signal peptidase complex subunit	Fast neutron microarray based cloning
(43)	nst1	NAC transcription factor	Forward screening and Tnt1 flanking region cloning
(44)	mtstp1	WRKY transcription factor	Forward screening and Tnt1 flanking region cloning
(75)	ccr1, ccr2	Cinnamoyl CoA Reductase	Tnt1 reverse
(76)	ugt73f3	Glucosyl transferase	Tnt1 reverse
(45)	fcl1	Class M KNOX	Fast neutron forward, map based cloning and Tnt1 reverse
(77)	rdn1	Uncharacterized plant family	Mapping

Reference	Mutant	Gene	Approach
(78)	*fta, ftc*	MtFTa, MtFTc, protein ligands	Tnt1 reverse
(48)	*vpy*	Vapyrin	Microarray based cloning, Tnt1 reverse
(49)	*lha*	CYP716A12, Cytochrome P450	Flanking sequence tagging and TILLING
(50)	*MtSGR*	Stay green gene	Tnt1 forward and flanking sequence cloning
(51)	*slm1*	Auxin efflux carrier protein	Tnt1 forward and flanking sequence cloning
(52)	*stf*	Stenofolia, WUSCHEL-like homeobox transcription factor	Tnt1 forward and flanking sequence cloning
(71)	*mate2*	MATE	Tnt1 reverse
(53)	*irg1*	Cys(2)His(2) zinc finger transcription factor	Tnt1 forward and flanking sequence cloning
(79)	*mtpar*	MYB transcription factor	Tnt1 reverse

Table 3. *Medicago* genes characterized using mutants.

7. RNAi and VIGS

Reverse genetics studies in *Medicago truncatula* did not only take advantage of the many mutant populations available but also of techniques based on post-transcriptional gene silencing (PTGS). In this case plants are transformed with a construct that will produce double-stranded RNAs that will guide sequence-specific mRNA degradation of the target gene. The phenotype of the transformed plants can gradually vary from wild type to knock-out thus many transformants are needed to obtain the desired effect. Mild effects can be beneficial in case of essential genes whose complete loss-of-function may cause lethal phenotypes. RNAi in *M.truncatula* has been extensively used to study gene function but it has not been a matter of a functional genomics approach as for *Arabidopsis* and the AGRIKOLA collection (80). Nevertheless many gene functions have been characterized exploiting RNAi. A list of gene function and *Medicago truncatula* physiology studies that used RNAi approaches is reported in Table 4.

Reference	Silenced gene	Phenotype
(75)	Lyk3	Marked reduction of nodulation when inoculated with Sm 2011ΔNodFE-GFP
(75)	Lyk4	Effect on infection thread morphology
(81)	CDPK1	Reduced root hair and root cell lengths. Diminution of both rhizobial and mycorrhizal symbiotic colonization.
(82)	DMI2	Reduction of organelle-like symbiosomes in nodules

Reference	Silenced gene	Phenotype
(83)	NFP	Nod-
(76)	MtHAP2-1	Alteration of nodule development
(84)	MtCPK3	Increased average nodule number
(85)	MtCRE1	Cytokinin-insensitive roots, increate number of lateral roots, strong reduction in nodulation.
(86)	MtPIN2, MtPIN3, MtPIN4	Reduced number of nodules
(87)	CHS	Reduced levels of flavonoids and subsequent inability to nodulate
(88)	PR10-1 (pathogenesis related)	Reduced colonization by the root pathogen A.euteiches.
(89)	HMGR1	Dramatic decrease in nodulation.
(90)	IPD3	No obvious phenotype observed
(63)	MtPT4	Premature death of mycorrhizal arbuscules.
(91)	MtSNF4b	Reduced seed longevity, alteration in non reducing sugar content.
(92)	ENOD40-1, ENOD40-2	Reduced nodule number and altered symbiosome development.
(93)	MtFNSII-1, MtFNSII-2	Reduced nodulation
(94)	MtCDD1	Alteration of the Arbuscular Mycorrhizal – mediated accumulation of apocarotenoids
(95)	MtDXS2	Reduction of AM-induced apocarotenoid accumulation.
(78)	MtSERF1	Strong inhibition of somatic embryogenesis
(73)	MtPI, MtNGL9	Altered flower development
(96)	MtWUS	Strong inhibition of somatic embryogenesis
(65)	Srlk	Transgenic root growth less inibited by salt stress.
(97)	FLOT2, FLOT4	Reduced nodulation and root development.
(98)	MtMSBP1	Aberrant mycorrhizal phenotype with thik and septated appressoria, decrease number of arbuscules and distorted arbuscule morphology.
(99)	MtCDC16	Decreased number of lateral roots and increased number of nodules. Reduced sensivity to auxin.
(100)	NPR1	Acceleration of root hair curling at the beginning of symbiosis establishment
(101)	MtSNARP2	Aberrant early senescent nodules where differentiated bacteroids degenerate rapidly.
(74)	MtSYMREM1	Reduced nodulation and abnormal nodule development

Reference	Silenced gene	Phenotype
(102)	MtN5	Reduced nodulation
(103)	Vapyrin	Impaired passage across epidermis by AM fungi. Abolition of arbuscule formation.
(104)	MtAOC1	No nodulation phenotype observed
(105)	γECS	Lower homoglutathione content. Lower biological nitrogen fixation associated with a reduction in the expression of the leghemoglobin and thioredoxin S1 genes. Reduction in nodule size.
(106)	MtSAP1	Lower level of storage globulin proteins, vicilin and legumin in seeds and germination deficiency.
(107)	MtNR1, MtNR2	Reduced nitrate or nitrite reductase activity and NO level.
(108)	MtNoa/Rif1	Decrease in NO production in roots but not in nodules. Reduction of nodule number and nitrogen fixation capacity.
(109)	MtROPGEF2	Effect on cytosolic Ca2+ gradient and subcellular structure of root hairs. Reduced root hair growth.
(110)	MtROP9	Reduced growth , no ROS generation after microbial infection. Promoted mycorrhizal and A.euteiches early hyphal root colonization. Impaired rhizobial colonization.
(111)	MtNAC969	Improved growth under salt stress.

Table 4. Use of RNAi approaches in *Medicago truncatula*.

Virus-induced gene silencing (VIGS) is a PTGS technique that can be used transiently by scrubbing leaves or introducing the viral vector in the plant by agro-infiltration. VIGS is being used for large scale forward genetics screening by inoculation of cDNA library and subsequent identification of the gene involved in the process of interest (112). Viral vectors working on *Medicago truncatula* have been recently described. Grønlund *et al.* used successfully a Pea Early Browning Virus (PEBV) based vector for both transient expression of reporter genes and for silencing of the Phytoene Desaturase (PDS) gene that causes a bleaching phenotype (113). Várallyay and colleagues constructed two VIGS vectors based on the Sunnhemp Mosaic Virus (SHMV) that can systemically infect *M.truncatula* without causing severe symptoms and reported a successful silencing of the Chlorata 42 gene (114). Large scale screenings based on VIGS analysis have not been reported for *M. truncatula* as far.

8. Perspectives

Functional genomics of forage legumes started with the aim of determining the molecular and genetic bases of nitrogen fixation and since the beginning mutant collections have been thoroughly screened also for mycorrhyzal symbiosis. These aspects are still being investigated and we expect that many more results will be published in the next years. A better understanding of nitrogen fixation and symbiosis is fundamental for the development of a sustainable agriculture aiming at a reduction of inputs and at maintaining soil fertility. Nitrogen (N) is one of the crucial nutrients for all organisms including plants. The doubling of world food production in the past four decades was contributed by a sevenfold increase of N fertilization (115). The anthropogenic N which is mostly lost to air, water and land affects climate, the chemistry of the atmosphere, and the composition and function of terrestrial and aquatic ecosystems (116). Improving the ability of plants to exploit environmental nitrogen would decrease N fertilization and its negative consequences; therefore a deep understanding of legume symbiosis with nitrogen fixing bacteria could help the long term goal of transferring the associative ability of legume species to non-symbiotic crops of agronomic relevance. As a consequence functional genomics of nodulation will have an impact on reduction of intensive agriculture practices with benefits for the preservation of environment and quality of human activities.

Another positive role for legumes in an environmental perspective is addressed by species such as *Lotus spp.* that have strong adaptive characteristics making them good candidates for restoration and phytoremediation of degraded environments (117). This happens in the Flooding Pampa (Argentina) where the presence of proteinaceous forages was re-established by the introduction of *L. tenuis*, being the other legume species reduced by the harsh environmental condition.

Pastures and feedstuff including forage legumes have a higher quality compared to those based only on grasses and provide an important input of protein in animal nutrition. More recently public and scientific debate has reassessed forage legumes importance for the quality of livestock nutrition and welfare has having relevant consequences on the quality of final products (meat, milk etc.) and ultimately on human health. This happened because of the occurrence of bovine spongiform encephalopathy (BSE) related to the traditional use of offal in animal feed lots as a source of protein.

Functional genomics in *M.truncatula* proved useful in the study and comprehension of many aspects of plant development and plant secondary metabolism that could not be discovered in earlier models such as *Arabidopsis*. The availability of genomics tools in an increasing number of species has the effect of widening the possibility of new discoveries in the field of plant biology. Worth mentioning the recent advances in understanding compound leaf development and zygomorphic flower ontogeny based on the analysis of several mutants in *M.truncatula* .

Living organisms, and among them plants, can be considered as an abundant and diverse set of biofactories with the ability to synthesize an enormous variety of chemical compounds. Legumes contain chemicals that can prove useful for their anti-oxidant, anti-viral, anti-microbial, anti-diabetic, anti-allergenic and anti-inflammatory properties (118) . These properties are related to secondary molecules such as flavonoids and saponins.

Modest levels of protoanthocyanidins (PAs) in forages reduce the occurrence of bloat and at the same time promote increased dietary protein nitrogen utilization in ruminant animals (119). The lack of PAs in the leaves of the major forage legume such as alfalfa has prompted studies for the understanding of the molecular and cellular biology of PA polymerization, transport, and storage helped by the functional genomics tools available for *M.truncatula*. Recent positive achievements were obtained by biotechnological strategies based on the overexpression of MYB transcription factors that induced PAs accumulation in both alfalfa and clover leaves (79).

In addition to well-known beneficial properties of flavonoids (cit) recent evidence suggests that flavonoids themselves, particularly fractions rich in PAs, can significantly reduce cognitive deterioration in animal model systems (120-122), and may more generally promote improvements in memory acquisition, consolidation, storage, and retrieval under nondegenerative conditions.

In Chinese medicine one of the oldest herbal medicine was obtained by the roots of the legume plant licorice (*Glychyrriza glabra*).containing the triterpenoid saponin glychyrrizin exhibiting a wide range of pharmacological activities. Cytochrome P450 monooxygenases were proved to be responsible for synthesis of glychyrrizin via oxidative steps based on biochemical experiments (123).

In forage legumes saponins can be toxic to monogastric animals and reduce forage palatability for ruminants. Mutant analysis in *M.truncatula* has unveiled the genetic control of key biosynthetic steps for saponins related to oxidation and glycosilation (49;124), opening possibilities of biotechnological manipulation of saponins in alfalfa.

Both human and animal nutritional science are bound to profit from plant genetic analysis and nutritional genomics, opening possibilities to more personalized approaches to medicine and improvement of the quality of life.

Author details

Francesco Panara and Ornella Calderini
CNR (National Council of Research) – Institute of Plant Genetics, Perugia, Italy

Andrea Porceddu*
University of Sassari, Italy

9. References

[1] Sato S, Sachiko I,T Satoshi . Structural analyses of the genomes in legumes. Curr Opin Plant Biol 2010;13 (2):146-52.
[2] Young ND, Debellé Fdr, Oldroyd GED, Geurts R, Cannon SB, Udvardi MK, et al. The Medicago genome provides insight into the evolution of rhizobial symbioses. Nature 2011;480 (7378):520-4.

---

* Corresponding Author

[3] Branca A, Paape TD, Zhou P, Briskine R, Farmer AD, Mudge J, et al. Whole-genome nucleotide diversity, recombination, and linkage disequilibrium in the model legume Medicago truncatula. Proc Natl Acad Sci U S A 2011;108 (42):E864-E870.

[4] Paape T, Zhou P, Branca A, Briskine R, Young N and Tiffin P. Fine-Scale Population Recombination Rates, Hotspots, and Correlates of Recombination in the Medicago truncatula Genome. Genome Biol Evol 2012;4 (5):726-37.

[5] Benaben V, Duc G, Lefebvre V. TE7, An Inefficient Symbiotic Mutant of Medicago truncatula Gaertn. cv Jemalong. Plant Physiol 1995;107 (1):53-62.

[6] Sagan M, Morandi D, Tarenghi E and Duc G.. Selection of nodulation and mycorrhizal mutants in the modelplantMedicagotruncatula (Gaertn.) after Î³-raymutagenesis. Plant Science 1995;111 :63-71.

[7] Cook DR, VandenBosch K, de Bruijn FJ. Model legumes get the nod. The Plant Cell 1997;3 :275.

[8] Penmetsa R and Cook DR. A Legume Ethylene-Insensitive Mutant Hyperinfected by Its Rhizobial Symbiont. Science 1997;275 (5299):527-30.

[9] Scholte M, d'Erfurth I, Rippa S, Mondy S, Cosson V, Durand P, et al. T-DNA tagging in the model legume Medicago truncatula allows efficient gene discovery. Molecular Breeding 2002;10 (4) :203-15.

[10] d'Erfurth I, Cosson V, Eschstruth A, Lucas H, Kondorosi A and Ratet P. Efficient transposition of the Tnt1 tobacco retrotransposon in the model legume Medicago truncatula. Plant J 2003;34 (1):95-106.

[11] Wang H. Fast neutron bombardment (FNB) mutagenesis for forward and reverse genetic studies in plants. ed. Global Science Books, Isleworth, UK, pp 629-639; 2006.

[12] Tadege M, Ratet P and Mysore K. Insertional mutagenesis: a Swiss Army knife for functional genomics of Medicago truncatula. Trends Plant Sci 2005;10 (5):229-35.

[13] Tadege M, Wen J, He J, Tu H, Kwak Y, Eschstruth A, et al. Large-scale insertional mutagenesis using the Tnt1 retrotransposon in the model legume Medicago truncatula. Plant J 2008;54 (2):335-47.

[14] Porceddu A, Panara F, Calderini O, Molinari L, Taviani P, Lanfaloni L, et al. An Italian functional genomic resource for Medicago truncatula. BMC Res Notes 2008;1 :129.

[15] Oldach KH, Peck DM, Cheon JUDY, Williams KJ and Nair R. Identification of a Chemically Induced Point Mutation Mediating Herbicide Tolerance in Annual Medics (Medicago spp.). Annals of Botany 2008;101 :997-1005.

[16] Iantcheva A, Chabaud M, Cosson V, Barascud M, Schutz B, Primard-Brisset C, et al. Osmotic shock improves Tnt1 transposition frequency in Medicago truncatula cv Jemalong during in vitro regeneration. Plant Cell Rep 2009;28 (10):1563-72.

[17] Rogers C, Wen J, Chen R and Oldroyd G. Deletion-Based Reverse Genetics in Medicago truncatula. Plant Physiology 2009;151(3) :1077-86.

[18] Signor CL, Savois V, Aubert Gg, Verdier J, Nicolas M, Pagny G, et al. Optimizing TILLING populations for reverse genetics in Medicago truncatula. Plant Biotechnol J 2009;7 (5):430-41.

[19] Sagan M, deLarambergue H. Genetic analysis of symbiosis mutants in Medicago truncatula. ed. Kluwer Academic Publishers, Dordrecht, The Netherlands; 1998.

[20] Penmetsa RV. Production and characterization of diverse developmental mutants of Medicago truncatula. Plant Physiol 2000;123 (4):1387-98.

[21] Catoira R, Timmers AC, Maillet F, Galera C, Penmetsa RV, Cook D, et al. The HCL gene of Medicago truncatula controls Rhizobium-induced root hair curling. Development 2001;128 (9):1507-18.

[22] Nakata PA. Isolation of Medicago truncatula mutants defective in calcium oxalate crystal formation. Plant Physiol 2000;124 (3):1097-104.

[23] Wais RJ, Galera C, Oldroyd G, Catoira R, Penmetsa RV, Cook D, et al. Genetic analysis of calcium spiking responses in nodulation mutants of Medicago truncatula. Proc Natl Acad Sci U S A 2000;97 (24):13407-12.

[24] Cohn JR, Uhm T, Ramu S, Nam YW, Kim DJ, Penmetsa RV, et al. Differential regulation of a family of apyrase genes from Medicago truncatula. Plant Physiol 2001;125 (4):2104-19.

[25] McConn MM. Oxalate reduces calcium availability in the pads of the prickly pear cactus through formation of calcium oxalate crystals. J Agric Food Chem 2004;52 (5):1371-4.

[26] Oldroyd G. Identification and Characterization of Nodulation-Signaling Pathway 2, a Gene of Medicago truncatula Involved in Nod Factor Signaling. Plant Physiology 2003;131(3) :1027-32.

[27] Amor BB, Shaw SL, Oldroyd GED, Maillet F, Penmetsa RV, Cook D, et al. The NFP locus of Medicago truncatula controls an early step of Nod factor signal transduction upstream of a rapid calcium flux and root hair deformation. Plant J 2003;34 (4):495-506.

[28] Penmetsa RV, Uribe P, Anderson J, Lichtenzveig J, Gish JC, Nam YW, et al. The Medicago truncatula ortholog of Arabidopsis EIN2, sickle, is a negative regulator of symbiotic and pathogenic microbial associations. Plant J 2008;55 (4):580-95.

[29] Kuppusamy KT, Endre G, Prabhu R, Penmetsa RV, Veereshlingam H, Cook DR, et al. LIN, a Medicago truncatula gene required for nodule differentiation and persistence of rhizobial infections. Plant Physiol 2004;136 (3):3682-91.

[30] Veereshlingam H, Haynes JG, Penmetsa RV, Cook DR, Sherrier DJ. nip, a symbiotic Medicago truncatula mutant that forms root nodules with aberrant infection threads and plant defense-like response. Plant Physiol 2004;136 (3):3692-702.

[31] Bright LJ, Liang Y, David, .Mitchell and Harris J. The LATD Gene of Medicago truncatula Is Required for Both Nodule and Root Development. MolecularPlant Microbe Interaction 2005;18(6) :521-432.

[32] Morandi D, Prado E, Sagan M&DGr. Characterisation of new symbiotic Medicago truncatula (Gaertn.) mutants, and phenotypic or genotypic complementary information on previously described mutants. Mycorrhiza 2005;15 (4):283-9.

[33] Starker CG, Parra-Colmenares AL, Smith L, Mitra RM. Nitrogen fixation mutants of Medicago truncatula fail to support plant and bacterial symbiotic gene expression. Plant Physiol 2006;140 (2):671-80.

[34] Middleton PH, Jakab J, Penmetsa RV, Starker CG, Doll J, Kalò P, et al. An ERF transcription factor in Medicago truncatula that is essential for Nod factor signal transduction. Plant Cell 2007;19 (4):1221-34.

[35] Wang H, Chen J, Wen J, Tadege M, Li G, Liu Y, et al. Control of compound leaf development by FLORICAULA/LEAFY ortholog SINGLE LEAFLET1 in Medicago truncatula. Plant Physiol 2008;146 (4):1759-72.

[36] Vernié T, Moreau S, de Billy Fo, Plet J, Combier JP, Rogers C, et al. EFD Is an ERF transcription factor involved in the control of nodule number and differentiation in Medicago truncatula. Plant Cell 2008;20 (10):2696-713.

[37] Teillet A, Garcia J, de Billy Fo, Gherardi Ml, Huguet T, Barker DG, et al. api, A novel Medicago truncatula symbiotic mutant impaired in nodule primordium invasion. Mol Plant Microbe Interact 2008;21 (5):535-46.

[38] Arrighi JFo, Godfroy O, de Billy Fo, Saurat O, Jauneau A&GC. The RPG gene of Medicago truncatula controls Rhizobium-directed polar growth during infection. Proc Natl Acad Sci U S A 2008;105 (28):9817-22.

[39] Morandi D, le Signor C, Gianinazzi-Pearson V and Duc G. A Medicago truncatula mutant hyper-responsive to mycorrhiza and defective for nodulation. Mycorrhiza 2009;19 :4635-441.

[40] Chen J, Yu J, Ge L, Wang H, Berbel A, Liu Y, et al. Control of dissected leaf morphology by a Cys(2)His(2) zinc finger transcription factor in the model legume Medicago truncatula. Proc Natl Acad Sci U S A 2010;107 (23):10754-9.

[41] Laffont C, Blanchet S, Lapierre C, Brocard L, Ratet P, Crespi M, et al. The compact root architecture1 gene regulates lignification, flavonoid production, and polar auxin transport in Medicago truncatula. Plant Physiol 2010;153 (4):1597-607.

[42] Vassileva V, Zehirov G, Ugrinova M. Variable leaf epidermal leaf morphology in Tnt1 insertional mutants of the model legume Medicago truncatula LEGUME MEDICAGO TRUNCATULA. Biotechnol\&Biotechnol Eq 2010;24(4) :2060-5.

[43] Zhao Q, Gallego-Giraldo L, Wang H, Zeng Y, Ding SY, Chen F&DR. An NAC transcription factor orchestrates multiple features of cell wall development in Medicago truncatula. Plant J 2010;63 (1):100-14.

[44] Wang D, Griffitts J, Starker C, Fedorova E, Limpens E, Ivanov S, et al. A nodule-specific protein secretory pathway required for nitrogen-fixing symbiosis. Science 2010;327 (5969):1126-9.

[45] Peng J, Yu J, ans Yingqing Guo ans Guangming Li HW, Bai G and and Chen R. Regulation of Compound Leaf Development in Medicago truncatula by Fused Compound Leaf1, a Class M KNOX Gene. The Plant Cell 2011;23 :3929-43.

[46] Schnabel E, Journet EP, de Carvalho-Niebel F, Duc G. and Frugoli J. The Medicago truncatula SUNN gene encodes a CLV1-like leucine-rich repeat receptor kinase that regulates nodule number and root length. Plant Mol Biol 2005;58 (6):809-22.

[47] Murray JD. Invasion by invitation: rhizobial infection in legumes. Mol Plant Microbe Interact 2011;24 (6):631-9.

[48] Murray JD, Muni RRD, Torres-Jerez I, Tang Y, Allen S, Andriankaja M, et al. Vapyrin, a gene essential for intracellular progression of arbuscular mycorrhizal symbiosis, is also essential for infection by rhizobia in the nodule symbiosis of Medicago truncatula. Plant J 2011;65 (2):244-52.

[49] Carelli M, Biazzi E, Panara F, Tava A, Scaramelli L, Porceddu A, et al. Medicago truncatula CYP716A12 is a multifunctional oxidase involved in the biosynthesis of hemolytic saponins. Plant Cell 2011;23 (8):3070-81.

[50] Zhou C, Han L, Pislariu C, Nakashima J, Fu C, Jiang Q, et al. From model to crop: functional analysis of a STAY-GREEN gene in the model legume Medicago truncatula and effective use of the gene for alfalfa improvement. Plant Physiol 2011;157 (3):1483-96.

[51] Zhou C, Han L, Hou C, Metelli A, Qi L, Tadege M, et al. Developmental analysis of a Medicago truncatula smooth leaf margin1 mutant reveals context-dependent effects on compound leaf development. Plant Cell 2011;23 (6):2106-24.

[52] Tadege M&Mysore K. Tnt1 retrotransposon tagging of STF in Medicago truncatula reveals tight coordination of metabolic, hormonal and developmental signals during leaf morphogenesis. Mob Genet Elements 2011;1 (4):301-3.

[53] Uppalapati SR, Ishiga Y, Doraiswamy V, Bedair M, Mittal S, Chen J, et al. Loss of abaxial leaf epicuticular wax in Medicago truncatula irg1/palm1 mutants results in reduced spore differentiation of anthracnose and nonhost rust pathogens. Plant Cell 2012;24 (1):353-70.

[54] Mitra RM, Gleason CA, Edwards A, Hadfield J, Downie JA, GED GO&SL. A Ca2+/calmodulin-dependent protein kinase required for symbiotic nodule development: Gene identification by transcript-based cloning. Science 2004;101(13 :4701.

[55] O'Malley R and Ecker J. Linking genotype to phenotype using the Arabidopsis unimutant collection. The Plant Journal 2010;61 (6) :928-40.

[56] Fukai E, Umehara Y, Sato S, Endo M, Kouchi H, Hayashi M, et al. Derepression of the plant Chromovirus LORE1 induces germline transposition in regenerated plants. PLoS Genet 2010;6 (3):e1000868.

[57] Fukai E, Soyano T, Umehara Y, Nakayama S, Hirakawa H, Tabata S, et al. Establishment of a Lotus japonicus gene tagging population using the exon-targeting endogenous retrotransposon LORE1. Plant J 2012;69 (4):720-30.

[58] Urbanski DF, Malolepszy A, Stougaard Jand S.Ugerrøj. Genome-wide LORE1 retrotransposon mutagenesis and high-throughput insertion detection in Lotus japonicus. Plant J 2012;69 (4):731-41.

[59] Rakocevic A, Mondy S, Tirichine Ll, Cosson V, Brocard L, Iantcheva A, et al. MERE1, a low-copy-number copia-type retroelement in Medicago truncatula active during tissue culture. Plant Physiol 2009;151 (3):1250-63.

[60] Ratet P. Medicago truncatula handbook. ed. Noble Foundation; 2006.

[61] Cheng X, Wen J, Tadege M, Ratet P. Reverse genetics in medicago truncatula using Tnt1 insertion mutants. Methods in molecular biology 2011;678 :179-190.

[62] Endre G, Kereszt A, Kevei Zn, Mihacea S, Kalò Pand Kiss G. A receptor kinase gene regulating symbiotic nodule development. Nature 2002;417 (6892):962-6.

[63] Ané JM, Kiss GrB, Riely BK, Penmetsa RV, Oldroyd G, Ayax C, et al. Medicago truncatula DMI1 required for bacterial and fungal symbioses in legumes. Science 2004;303 (5662):1364-7.

[64] Lèvy J, Bres C, Geurts R, Chalhoub B, Kulikova O, Duc G, et al. A putative Ca2+ and calmodulin-dependent protein kinase required for bacterial and fungal symbioses. Science 2004;303 (5662):1361-4.

[65] Kalò P, Gleason C, Edwards A, Marsh J, Mitra RM, Hirsch S, et al. Nodulation signaling in legumes requires NSP2, a member of the GRAS family of transcriptional regulators. Science 2005;308 (5729):1786-9.

[66] Smit P, Raedts J, Portyanko V, Debellé Fdr, Gough C, Bisseling T and Geurts R. NSP1 of the GRAS protein family is essential for rhizobial Nod factor-induced transcription. Science 2005;308 (5729):1789-91.

[67] Benlloch R, d'Erfurth I, Ferrandiz C, Cosson V, Beltràn JP, Canas LA, et al. Isolation of mtpim proves Tnt1 a useful reverse genetics tool in Medicago truncatula and uncovers new aspects of AP1-like functions in legumes. Plant Physiol 2006;142 (3):972-83.

[68] Javot H, Penmetsa RV, Terzaghi N, Cook DR. A Medicago truncatula phosphate transporter indispensable for the arbuscular mycorrhizal symbiosis. Proc Natl Acad Sci U S A 2007;104 (5):1720-5.

[69] Kiss E, Olàh Br, Kalò P, Morales M, Heckmann AB, Borbola A, et al. LIN, a novel type of U-box/WD40 protein, controls early infection by rhizobia in legumes. Plant Physiol 2009;151 (3):1239-49.

[70] de Lorenzo L, Merchan F, Laporte P, Thompson R, Clarke J, Sousa C and Crespi M. A novel plant leucine-rich repeat receptor kinase regulates the response of Medicago truncatula roots to salt stress. Plant Cell 2009;21 (2):668-80.

[71] Zhao J and Dixon R. MATE transporters facilitate vacuolar uptake of epicatechin 3'-O-glucoside for proanthocyanidin biosynthesis in Medicago truncatula and Arabidopsis. Plant Cell 2009;21 (8):2323-40.

[72] Peel GJ, Pang Y, Modolo LV. The LAP1 MYB transcription factor orchestrates anthocyanidin biosynthesis and glycosylation in Medicago. Plant J 2009;59 (1):136-49.

[73] Benlloch R, Roque En, Ferràndiz C, Cosson V, Caballero T, Penmetsa RV, et al. Analysis of B function in legumes: PISTILLATA proteins do not require the PI motif for floral organ development in Medicago truncatula. Plant J 2009;60 (1):102-11.

[74] Lefebvre B, Timmers T, Mbengue M, Moreau S, Hervé C, Tòth K, et al. A remorin protein interacts with symbiotic receptors and regulates bacterial infection. Proc Natl Acad Sci U S A 2010;107 (5):2343-8.

[75] Zhou R, Jackson L, Shadle G, Nakashima J, Temple S, Chen F. Distinct cinnamoyl CoA reductases involved in parallel routes to lignin in Medicago truncatula. Proceedings of the National Academy of Sciences National Acad Sciences; 2010;107(41):17803-17808.

[76] Naoumkina M and Dixon R. Characterization of the mannan synthase promoter from guar (Cyamopsis tetragonoloba). Plant Cell Rep 2011;30 (6):997-1006.

[77] Schnabel EL, Kassaw TK, Smith LS, Marsh JF, Oldroyd GE, Long SR. The ROOT DETERMINED NODULATION1 gene regulates nodule number in roots of Medicago truncatula and defines a highly conserved, uncharacterized plant gene family. Plant Physiol 2011;157 (1):328-40.

[78] Laurie ÌR, Diwadkar P, Jaudal M, Zhang L, Hecht V, Wen J, et al. The Medicago FLOWERING LOCUS T Homolog, MtFTa1, Is a Key Regulator of Flowering Time. Plant Physiology 2011;156 :2207-24.

[79] Verdier J, Zhao J, Torres-Jerez I, Ge S, Liu C, He X, et al. MtPAR MYB transcription factor acts as an on switch for proanthocyanidin biosynthesis in Medicago truncatula. Proc Natl Acad Sci U S A 2012;109 (5):1766-71.

[80] Hilson Pierre, Small Ian, Kuiper Martin. European consortia building reference resources for Arabidopsis functional genomics. Curr Opin Plant Biol 2012;6:426-9.

[81] Ivashuta S, Liu J, Liu J, Lohar DP, Haridas S, Bucciarelli B, et al. RNA interference identifies a calcium-dependent protein kinase involved in Medicago truncatula root development. Plant Cell 2005;17 (11):2911-21.

[82] Limpens E, Mirabella R, Fedorova E, Franken C, Franssen H, Bisseling T&GR. Formation of organelle-like N2-fixing symbiosomes in legume root nodules is controlled by DMI2. Proc Natl Acad Sci U S A 2005;102 (29):10375-80.

[83] Arrighi JF, Barre A, Amor BB, Bersoult A, Soriano LC, Mirabella R, et al. The Medicago truncatula lysin [corrected] motif-receptor-like kinase gene family includes NFP and new nodule-expressed genes. Plant Physiol 2006;142 (1):265-79.

[84] Gargantini PR, Gonzalez-Rizzo S, Chinchilla D, Raices M, Giammaria V, Ulloa RM, et al. A CDPK isoform participates in the regulation of nodule number in Medicago truncatula. Plant J 2006;48 (6):843-56.

[85] Gonzalez-Rizzo S, Crespi M and Frugier F. The Medicago truncatula CRE1 cytokinin receptor regulates lateral root development and early symbiotic interaction with Sinorhizobium meliloti. Plant Cell 2006;18 (10):2680-93.

[86] Huo X, Schnabel E, Hughes K and Frugoli J. RNAi Phenotypes and the Localization of a Protein::GUS Fusion Imply a Role for Medicago truncatula PIN Genes in Nodulation. J Plant Growth Regul 2006;25 (2):156-65.

[87] Wasson AP, Pellerone FI. Silencing the flavonoid pathway in Medicago truncatula inhibits root nodule formation and prevents auxin transport regulation by rhizobia. Plant Cell 2006;18 (7):1617-29.

[88] Colditz F, Niehaus K and Krajinski F. Silencing of PR-10-like proteins in Medicago truncatula results in an antagonistic induction of other PR proteins and in an increased tolerance upon infection with the oomycete Aphanomyces euteiches. Planta 2007;226 (1):57-71.

[89] Kevei Zn, Lougnon G, Mergaert P, Horvàth GbV, Kereszt A, Jayaraman D, et al. 3-hydroxy-3-methylglutaryl coenzyme a reductase 1 interacts with NORK and is crucial for nodulation in Medicago truncatula. Plant Cell 2007;19 (12):3974-89.

[90] Messinese E, Mun JH, Yeun LH, Jayaraman D, Rougé P, Barre A, et al. A novel nuclear protein interacts with the symbiotic DMI3 calcium- and calmodulin-dependent protein kinase of Medicago truncatula. Mol Plant Microbe Interact 2007;20 (8):912-21.

[91] Rosnoblet C, Aubry C, Leprince O, Vu BL, Rogniaux H and Buitink J. The regulatory gamma subunit SNF4b of the sucrose non-fermenting-related kinase complex is involved in longevity and stachyose accumulation during maturation of Medicago truncatula seeds. Plant J 2007;51 (1):47-59.

[92] Wan X, Hontelez J, Lillo A, Guarnerio C, van de Peut D, Fedorova E, et al. Medicago truncatula ENOD40-1 and ENOD40-2 are both involved in nodule initiation and bacteroid development. J Exp Bot 2007;58 (8):2033-41.

[93] Zhang J, Subramanian S, Zhang Y&Yu O. Flavone synthases from Medicago truncatula are flavanone-2-hydroxylases and are important for nodulation. Plant Physiol 2007;144 (2):741-51.

[94] Floss DS, Hause B, Lange PR, Kùster H, Strack D and Walter M. Knock-down of the MEP pathway isogene 1-deoxy-D-xylulose 5-phosphate synthase 2 inhibits formation of arbuscular mycorrhiza-induced apocarotenoids, and abolishes normal expression of mycorrhiza-specific plant marker genes. Plant J 2008;56 (1):86-100.

[95] Floss DS, Schliemann W, Schmidt JÃ, Strack D and Walter M. RNA interference-mediated repression of MtCCD1 in mycorrhizal roots of Medicago truncatula causes accumulation of C27 apocarotenoids, shedding light on the functional role of CCD1. Plant Physiol 2008;148 (3):1267-82.

[96] Chen SK, Kurdyukov S, Kereszt A, Wang XD, Gresshoff PM. The association of homeobox gene expression with stem cell formation and morphogenesis in cultured Medicago truncatula. Planta 2009;230 (4):827-40.

[97] Haney CH. Plant flotillins are required for infection by nitrogen-fixing bacteria. Proc Natl Acad Sci U S A 2010;107 (1):478-83.

[98] Kuhn H, Kùster H&RN. Membrane steroid-binding protein 1 induced by a diffusible fungal signal is critical for mycorrhization in Medicago truncatula. New Phytol 2010;185 (3):716-33.

[99] Kuppusamy KT, Ivashuta S, Bucciarelli B, Vance CP, Gantt JS. Knockdown of CELL DIVISION CYCLE16 reveals an inverse relationship between lateral root and nodule numbers and a link to auxin in Medicago truncatula. Plant Physiol 2009;151 (3):1155-66.

[100] Peleg-Grossman S, Golani Y, Kaye Y, Melamed-Book N and Levine A. NPR1 protein regulates pathogenic and symbiotic interactions between Rhizobium and legumes and non-legumes. PLoS One 2009;4 (12):e8399.

[101] Laporte P, Satiat-Jeunemaìtre B, Velasco I, Csorba T, Van de Velde W, Campalans A, et al. A novel RNA-binding peptide regulates the establishment of the Medicago truncatula-Sinorhizobium meliloti nitrogen-fixing symbiosis. Plant J 2010;62 (1):24-38.

[102] Pii Y, Astegno A, Peroni E, Zaccardelli M, Pandolfini T and Crimi M. The Medicago truncatula N5 gene encoding a root-specific lipid transfer protein is required for the symbiotic interaction with Sinorhizobium meliloti. Mol Plant Microbe Interact 2009;22 (12):1577-87.

[103] Pumplin N, Mondo SJ, Topp S, Starker CG, Gantt JS. Medicago truncatula Vapyrin is a novel protein required for arbuscular mycorrhizal symbiosis. Plant J 2010;61 (3):482-94.

[104] Zdyb A, Demchenko K, Heumann J, Mrosk C, Grzeganek P, Gòbel C, et al. Jasmonate biosynthesis in legume and actinorhizal nodules. New Phytol 2011;189 (2):568-79.

[105] Msehli SE, Lambert A, Baldacci-Cresp F, Hopkins J, Boncompagni E, Smiti SA, et al. Crucial role of (homo)glutathione in nitrogen fixation in Medicago truncatula nodules. New Phytol 2011;192 (2):496-506.

[106] Gimeno-Gilles C, Gervais ML, Planchet E, Satour P, Limami AM. A stress-associated protein containing A20/AN1 zing-finger domains expressed in Medicago truncatula seeds. Plant Physiol Biochem 2011;49 (3):303-10.

[107] Horchani F, Prèvot M, Boscari A, Evangelisti E, Meilhoc E, Bruand C, et al. Both plant and bacterial nitrate reductases contribute to nitric oxide production in Medicago truncatula nitrogen-fixing nodules. Plant Physiol 2011;155 (2):1023-36.

[108] Pauly N, Ferrari C, Andrio E, Marino D, Piardi Sp, Brouquisse R, et al. MtNOA1/RIF1 modulates Medicago truncatula-Sinorhizobium meliloti nodule development without affecting its nitric oxide content. J Exp Bot 2011;62 (3):939-48.

[109] Riely BK, He H, Venkateshwaran M, Sarma B, Schraiber J, Anè J-M&CD. Identification of legume RopGEF gene families and characterization of a Medicago truncatula RopGEF mediating polar growth of root hairs. Plant J 2011;65 (2):230-43.

[110] Kiirika LM, Bergmann HF, Schikowsky C, Wimmer D, Korte J, Schmitz U, et al. Silencing of the Rac1 GTPase MtROP9 in Medicago truncatula Stimulates Early Mycorrhizal and Oomycete Root Colonizations But Negatively Affects Rhizobial Infection. Plant Physiol 2012;159 (1):501-16.

[111] de Zélicourt A, Diet A, Marion J, Laffont C, Ariel F, Moison Ml, et al. Dual involvement of a Medicago truncatula NAC transcription factor in root abiotic stress response and symbiotic nodule senescence. Plant J 2012;70 (2):220-30.

[112] Senthil-Kumar M and Mysore K. New dimensions for VIGS in plant functional genomics. Trends Plant Sci 2011;16 (12):656-65.

[113] Grønlund M, Constantin G, Piednoir E, Kovacev J, Johansen IE. Virus-induced gene silencing in Medicago truncatula and Lathyrus odorata. Virus Res 2008;135 (2):345-9.

[114] Vàrallyay E, Lichner Z, SÃ¡frÃ¡ny J, Havelda Z, Salamon P, Bisztray G&Bn. Development of a virus induced gene silencing vector from a legumes infecting tobamovirus. Acta Biol Hung 2010;61 (4):457-69.

[115] Ollivier J, Töwe S, Bannert A, Hai B, Kastl EM, Meyer A, et al. Nitrogen turnover in soil and global change. FEMS Microbiol Ecol 2011;78 (1):3-16.

[116] Galloway JN, Townsend AR, Erisman JW, Bekunda M, Cai Z, Freney JR, et al. Transformation of the nitrogen cycle: recent trends, questions, and potential solutions. Science 2008;320 (5878):889-92.

[117] Escaray FJ, Menendez AB, Gárriz A, Pieckenstain FL, Estrella MJ, Castagno LN, et al. Ecological and agronomic importance of the plant genus Lotus. Its application in grassland sustainability and the amelioration of constrained and contaminated soils. Plant Sci 2012;182 :121-33.

[118] Howieson JG. Nitrogen-fixing leguminous symbioses. %P ed. Springer; 2008.

[119] Dixon RA. Flavonoids and isoflavonoids: from plant biology to agriculture and neuroscience. Plant Physiol 2010;154 (2):453-7.

[120] Ho L, Chen LH, Wang J, Zhao W, Talcott ST, Ono K, et al. Heterogeneity in red wine polyphenolic contents differentially influences Alzheimer's disease-type neuropathology and cognitive deterioration. J Alzheimers Dis 2009;16 (1):59-72.

[121] Pasinetti GM, Zhao Z, Qin W, Ho L, Shrishailam Y, Macgrogan D, et al. Caloric intake and Alzheimer's disease. Experimental approaches and therapeutic implications. Interdiscip Top Gerontol 2007;35 :159-75.

[122] Wang J, Ferruzzi MG, Ho L, Blount J, Janle EM, Gong B, et al. Brain-targeted proanthocyanidin metabolites for Alzheimer's disease treatment. J Neurosci 2012;32 (15):5144-50.

[123] Seki A, Satoru S; Kiyoshi O; Masaharu M; Toshiyuki O; Hiroshi M; et al.. Triterpene functional genomics in licorice for identification of CYP72A154 involved in the biosynthesis of glycyrrhizin. Plant Cell 2011, 23(6) 4112-4123.

[124] Naoumkina MA, Modolo LV, Huhman DV, Urbanczyk-Wochniak E, Tang Y, Sumner LW. Genomic and coexpression analyses predict multiple genes involved in triterpene saponin biosynthesis in Medicago truncatula. Plant Cell 2010;22 (3):850-66.

Chapter 9

Dynamic Proteomics: Methodologies and Analysis

Sara ten Have, Kelly Hodge and Angus I. Lamond

Additional information is available at the end of the chapter

1. Introduction

Proteins are dynamic and any detailed description of the proteome must reflect the dynamic variations in protein properties. For example, most proteins form complexes with other protein partners, can undergo various post translational modifications and can accumulate in different sub compartments of the cell. Spatial and temporal variations between proteins in different compartments and/or cell types mean that each experiment for mass spectrometric analysis must be carefully designed to optimise the data that can be obtained. Recent improvements in experimental methodologies and in the resolution and sensitivity of Mass Spectrometers, have expanded the complexity of proteomic analysis that is now possible[1]. In this chapter we outline current workflows and methodologies that facilitate complex proteomic analyses, from the design and execution of experiments, though to the analysis and interpretation of the resulting data.

SILAC labelling can be used to quantitate a wide range of biological experiments based upon differential comparisons of two or three cell states or conditions. For example, immuno-precipitation and protein-protein interaction analysis, cellular fractionation for localisation studies and measurements of protein synthesis, degradation and turnover can all be quantitated using the SILAC approach [2-7].The SILAC approach can also be used to carry out high throughput analyses on entire proteomes and can help to identify subsets of proteins that respond to specific cellular perturbations.

Reliable interpretation of SILAC data requires computational analysis. Widely accessible spread sheet applications like excel are commonly used for this task. This involves numerous peptide and protein identifications, with several isotope ratio and/or intensity values associated with each identification. The interpretation of these data is often the most complex part of the proteomics experiment. How to go about data quality assurance and culling, as well as modelling the data in such a way as to draw valid conclusions will be discussed in this chapter.

Mass Spectrometry-based proteomics is evolving into a multidimensional analysis world (e.g. identification, quantification, space and time), where not only do we identify and quantify the proteome but also characterize changes in protein properties (e.g. subcellular location) at different time points, and under different conditions (e.g. response to a drug treatment). These types of analyses help to provide a functional characterisation of the genome and may facilitate the application of proteomics for clinical studies.

2. Mass Spectrometry of proteins

Mass Spectrometry of proteins is based on several principals of chemistry and physics, namely mass and generation of charged molecules, or ions. Given the known composition of amino acids (Figure 1), and the inferred knowledge of protein composition (from the Human Genome Project [8], the protein products are predicted from the genetic code), we can therefore compute in silico the predicted molecular weight of every protein. Additionally, the mass change resulting from any modification (made in the laboratory, for example reduction and alkylation of cysteine di-sulphide bonds) can be accurately predicted and therefore matches can be made between these calculated values and the experimental ion masses measured in a mass spectrometer.

For complex protein mixtures (e.g., cellular lysates, immune-precipitates, whole organism and/or tissue lysates) protein analysis is typically performed using the following methodology:

Protein solubilisation; A separation step to fractionate the complex protein mixture (e.g. gel electrophoresis, isoelectric focussing or size exclusion chromatography); Reduction and Alkylation (to disrupt di-sulphide bonds in proteins, and add a carbomidomethyl modification to cysteine residues to inhibit di-sulphide re-formation); Digestion using a proteolytic enzyme such as trypsin (as bottom up –or peptide level analysis is the most common form of protein analysis in mass spectrometry); HPLC reversed phase chromatography (which reduces the complexity of peptide samples sufficiently for the instrument to measure the individual ions); Electro-spray ionisation and tandem mass spectrometry (Figure 2). Through the process of electro-spray ionisation peptides can be charged (mostly positively charged), and the charge used to control their movement through the instrument. The mass spectrometer then performs a survey scan (characterising all of the peptide ion masses present in a given time window), followed by several sequencing scans, which isolate and fragment peptides, one ion mass at a time, by colliding each selected peptide ion with inert gas molecules, thereby generating fragment or daughter ions. These fragment ions characterise the amino acid sequence of the selected peptide. Using this analysis strategy the current generation of mass spectrometer can generate ~30,000 or more spectra from a typical protein lysate, thereby identifying and quantifying hundreds to thousands of separate proteins, depending on the complexity of the sample.

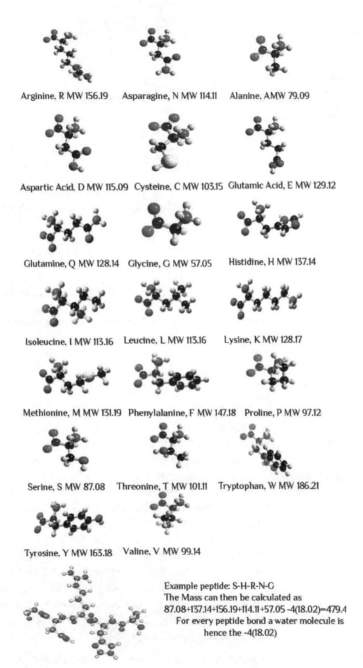

Arginine, R MW 156.19   Asparagine, N MW 114.11   Alanine, A MW 79.09

Aspartic Acid, D MW 115.09   Cysteine, C MW 103.15   Glutamic Acid, E MW 129.12

Glutamine, Q MW 128.14   Glycine, G MW 57.05   Histidine, H MW 137.14

Isoleucine, I MW 113.16   Leucine, L MW 113.16   Lysine, K MW 128.17

Methionine, M MW 131.19   Phenylalanine, F MW 147.18   Proline, P MW 97.12

Serine, S MW 87.08   Threonine, T MW 101.11   Tryptophan, W MW 186.21

Tyrosine, Y MW 163.18   Valine, V MW 99.14

Example peptide: S-H-R-N-G
The Mass can then be calculated as
87.08+137.14+156.19+114.11+57.05 -4(18.02)=479.4
For every peptide bond a water molecule is
hence the -4(18.02)

Figure 1. Structures of the common amino acids.

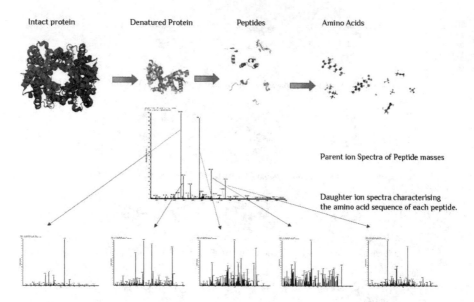

Figure 2. The basic workflow of bottom up proteomics, and how peptides are measured in a mass spectrometer.

3. SILAC

Relative quantification of proteins in two or more samples is aided by using isotope labelling techniques such as SILAC [9]. SILAC (Stable Isotope Labelling of Amino acids in Cell culture) is a quantitative method of analysis where specific amino acids (typically arginine and/or lysine) undergo a forced enrichment of heavy carbon, nitrogen and hydrogen isotopes (namely C13, N15 and deuterium, all of which are not radioactive) in cell culture, by using amino acid depleted media. After approximately 6 cell division cycles the vast majority of the proteins are completely substituted with the heavy isotope labelled amino acids (Figure 3) [9]. These basic amino acids are cleavage sites for the enzyme trypsin, ensuring every tryptic peptide measured contains a single SILAC isotope label. Modern mass spectrometers are extremely sensitive instruments that can detect the changes in weight caused by the presence of different isotopes e.g. 'medium' labelling of proteins is generally created by using L-arginine-$^{13}C_6$ $^{14}N_4$ and L-lysine $^2H_4$ (R6K4). Thus the 'medium' labelled arginine and lysine will have an increased mass of 6Da and 4Da, respectively, relative to the normal 'light' isotopes in naturally occurring amino acids. The MS spectra display these differences as distinctive double (for 2 SILAC labels) or triple (for 3 SILAC labels) peaks at a given mass for the endogenous/light peptide

Using this technology experimental scenarios have been established allowing the characterisation of the dynamic proteome [2-4, 6, 7].

SILAC specific instructions and some product information can be found in the Supplementary Methods section under 'SILAC'. Please note other suppliers are available and these can be found on the internet.

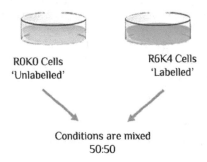

ROKO Cells
'Unlabelled'

R6K4 Cells
'Labelled'

Conditions are mixed
50:50

Each peptide yeilds a spectra with 2 peaks
which can be quantitatively measured.

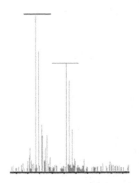

*'Light labelled' proteins should be prepared from cells grown on media identical to the heavy labelled media (i.e. media, to which amino acids with the normal 'light' isotopes and dialysed Calf serum are added). [Using 'normal' non-SILAC media is not sufficient as this will have a different composition, due mostly to the non-dialysed calf serum used in typical media. Non-dialysed calf serum may have more small molecules than dialysed, which could potentially change the growth rate of cells, therefore giving differential growth conditions between the control sample and the sample of interest.

Figure 3. The principle of SILAC. Cells which have been grown for >6 generations in SILAC media contain proteins completely substituted with heavy isotope-labelled amino acids. These different mass labels can be used for distinguishing and comparing proteins in a wide range of experimental conditions. When mixed with a control sample, with a different SILAC label, the resulting spectra of each peptide, allow accurate relative quantitation.

4. Protein interaction proteomics

As mentioned in the introduction, many proteins do not act in isolation, but form complexes with partner proteins and a major goal is to identify the detailed composition of these respective multi-protein complexes. However, the dynamic nature of the proteome means that there may not be a unique description of protein complexes. For example, at different cell cycle stages and/or under changed conditions (e.g. following drug treatment) the partner proteins in a complex might either change, or vary in abundance and/or modification state. Our aim, therefore, is to analyse both the composition and dynamic nature of protein complexes.

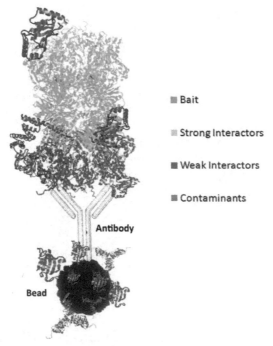

■ Bait

▨ Strong Interactors

■ Weak Interactors

■ Contaminants

Figure 4. Immunne-precipitation. The beads used in immune-precipitation experiments contribute a significant amount of non-specific binding proteins, and therefore need to be accounted for with the use of a bead control. This is done using cell lysate and the beads of choice, without the antibody, leading to a sample which contains only background proteins. With this information the genuine interactors can more easily be determined.

Using immune-precipitation (harnessing the specificity of antibodies to protein targets) for protein interaction experiments has long been the gold standard, particularly in combination with traditional western blot analysis. With the application of mass spectrometry to characterise immuno-precipitates, the analysis has now expanded to identifying hundreds of proteins in each IP. A large percentage of the proteins pulled down in an immune-precipitation experiment bind non-specifically, for example binding to the beads used as the solid substrate for the antibody rather than to the bait or target protein (Figure 4). The beads often have a high general binding affinity for protein [4, 10]. Without good controls these non-specifically binding proteins can occlude identification of the specific proteins of interest. To allow for this a bead control is often included as part of the experiment. A bead control is provided by applying equal amounts of the cell lysate of choice to the beads being used for the immune-precipitation, sans antibody. This generates a sample that will predominantly contain non-specific binding proteins, which can be identified during the analysis and distinguished from the genuine protein interaction partners in the complex of interest. While label free MS analysis is effective for this protocol, differential SILAC labelling to distinguish the control and experimental conditions (e.g. R0K0- bead control, R6K4 protein of interest, R10K8 protein + drug) improves the accuracy of quantitation and efficiency for this kind of analysis.

Extensive analysis of hundreds of immune-precipitations, with lysates from various different cell lines and bead types has been used to generate a database recording protein identification frequencies, i.e., recording the number of previous experiments where any given protein was identified [4]. The higher the number of times a protein is identified in different IP experiments, the more likely it represents a non-specific binder. This protein frequency library information is available as an online resource for comparing immune-precipitation results; see http://www.peptracker.com/datavisual/.

A basic immune-precipitation protocol can be found in the Supplementary Methods section under 'Immuno-precipitation Protocol'.

5. Dynamic proteomics: How it's done

The use of SILAC labelling enables a wide range of assay formats to be designed for quantitative comparison of protein properties under different conditions. For example, using SILAC in conjunction with cellular fractionation, immune-precipitation and time course experiments it is possible to analyse the kinetics of protein transport, synthesis, degradation and interaction.

Using a combination of physical and chemical separation methods, including differential density centrifugation, it is possible to fractionate cells and isolate subcellular organelles and components such as cytoplasm, nucleoplasm, membranes etc. There are also commercial kits available that can be used to fractionate cells and combined with MS analysis. The cellular fractionation most commonly used in our hands concentrates on distinguishing between cytoplasmic and nuclear localisation of proteins in eukaryotic cells, allowing analysis of compartmentalisation of protein function and nucleo-cytoplasmic transport under different cell growth conditions and responses [3, 11]. Figure 5 illustrates the procedure and the specifics of the methodology can be found in the Supplementary Methods under 'Cellular Fractionation'.

The principles of the fractionation strategy, as applied to mammalian cells grown in culture, are as follows; application of a hypotonic (low salt) buffer to freshly trypsinised cells, followed by a gentle mechanical disruption with a dounce homogeniser. This causes the cells to swell, and hence disrupts the outer cell membrane. The resulting 'cellular' suspension is centrifuged such that larger organelles, including the nucleus (which at this stage is intact) will spin down into a pellet, whilst the soluble material and smaller cytoplasmic material will stay in the supernatant. Thereafter stronger mechanical disruption is employed (e.g. sonication) to lyse the nucleus, and one or more additional fractionation steps (e.g. density gradients) are used to separate organelles and other subcellular structures based on properties such as their size, density and/or shape.

This can be combined with MS-based approaches and SILAC to determine changes in the subcellular organisation of the proteome induced by stress or other perturbations (e.g. UV, drug treatment etc.). This is done by growing cells in media with different SILAC labels, using one of the labels as an untreated control sample (e.g. 'light') while exposing cells grown in a different label (e.g. 'medium' or 'heavy') to the perturbation, e.g. stress, drug treatment. After incubating for the desired time, which will vary depending on the treatment being performed, equal numbers of cells from each control and experimental sample can be mixed and the fractionation protocol carried out. Alternatively, the cell fractionation can be performed separately for the different samples and then mixed to combine equal amounts of protein from each (Figure 5). In this technique proteins remaining unchanged as a result of the perturbation will show a SILAC ratio for the control and experimental isotopic forms of ~1 (or if a log ratio is plotted, 0). In contrast, proteins which have been altered as a result of the experimental treatment (e.g. moved from cytoplasm to nucleus) will show either an increased or decreased SILAC ratio, according to the design of the experiment. This conveniently highlights a particular subset of proteins that may respond to a specific perturbation and provides in parallel a direct comparison with the bulk response of the large number of cell proteins sampled in high throughput.

This approach can also be used in combination with a Pulse SILAC experimental set-up, as discussed below.

The cellular fractionation protocol described above allows the characterisation of changes in the steady state localisation of proteins and of kinetics of protein movement, but this is not the full story. Although the location of a protein is fundamental to its function, the change induced by your experimental variable might also affect protein turnover, either by changing rates of protein synthesis, degradation or both. So how do we characterise this? Pulse SILAC techniques have enabled an elegant experimental procedure to characterise the time dynamics of the proteome [3, 12, 13]. This involves generating a population of completely labelled cells in medium label (e.g.R6K4), and switching the media over to heavy (e.g.R10K8). Over time conversion of all the medium labelled protein into heavy labelled protein occurs. Collecting cells at various time points, and mixing these with light labelled cells (50:50 as per usual SILAC) as a control steady state of protein expression (Figure 6), gives samples which characterise protein synthesis and degradation (13).

The benefit of this kind of experimental set up is evident in the downstream data analysis. Decrease in the medium to light ratios describes the degradation rate of a given protein, whilst increase in the heavy to light ratio describes the synthesis of new proteins. The time point at which these 2 curves intersect (assuming you have a sufficient number of time points for accurate measurements) describes the time required for turnover of 50% of the protein. Analysis of proteome turnover in the HeLa and HCT116 cell lines has been carried out and made publicly available at http://www.peptracker.com/turnoverInformation/.

Bearing in mind different cell lines have varying cell cycle length, the online Protein Turnover Viewer can allow comparison of new results with this database, and hence reveal differences in behaviour between cell lines. The Protein Turnover Viewer has an easily navigable interface, allowing Uniprot identifiers to be used to identify a protein of interest to find out the data on its turnover.

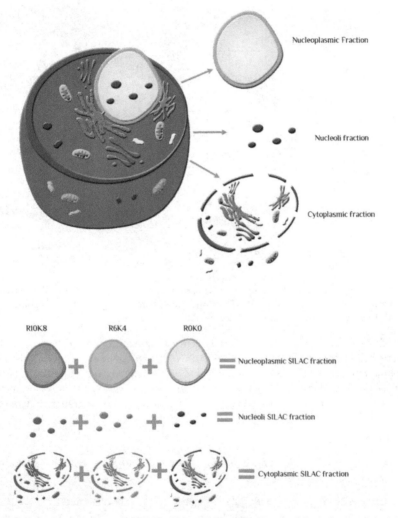

Figure 5. Cellular fractionation, and SILAC cellular fractionation. The physiological properties of the cellular structure enable effective separation of parts of the cell, using combinations of chemistry, centrifugal properties and varying strengths of mechanical disruption. This method in combination with SILAC enables characterisation of different conditions in one experiment, describing quantitatively the regulation and location of proteins.

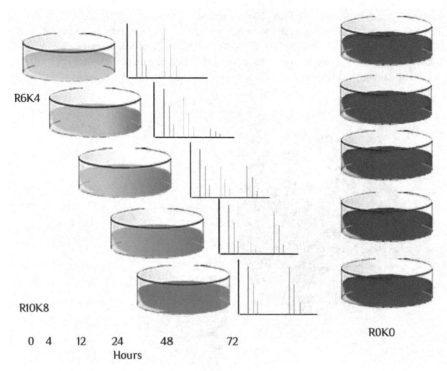

R6K4

R10K8

0   4   12   24   48   72

Hours

R0K0

Figure 6. Pulse SILAC. Pulse SILAC uses an established labelled population of cells and when a media swap (i.e. from R6K4 to R10K8) is instigated, measurements of protein degradation and synthesis can be performed, when mixed with a control population (R0K0).

This technique is not only useful for steady state, or 'normal' protein turnover analysis. It fits very well to drug treatment kinetics, microRNA effects, DNA damage analysis (e.g. UV or chemical induced), or physiological perturbations (e.g. hypoxia or other forms of stress). Analysis of the resulting data is more complex than a more simple SILAC experiment and the data set larger, but provides a useful wealth of information about protein dynamics.

6. Data analysis

Data analysis of SILAC experiments needs to be tailored to the specific question, but the beginnings of the analysis process are very similar and can follow this method:

MaxQuant → Data Culling → Population Statistics → Data Grouping

6.1. MaxQuant

MaxQuant is a comprehensive software package widely used for the analysis and quantitation of MS-based proteomic data, including SILAC, that was created by Jurgen

Cox and Matthias Mann [14, 15]. It is made available as freeware and can be downloaded from http://maxquant.org/. MaxQuant includes a search engine that can use raw MS data from the mass spectrometer, perform peak picking, mass recalibration, SILAC pair matching and quantification, label free quantification, database searching (using Andromeda), and output peptide and protein data in extensive detail [14, 15]. While other commercial and freeware software options are also available for analysis of MS data we routinely use the MaxQuant package which works very well specifically for the protocols described here.

6.2. Data grouping

Data grouping is a way of making large data sets easier to manage. In an ideal world having a database with experimental values linked to reliable meta data describing the experimental parameters is the best case scenario for proteomic data management [2-4, 15].

Online versions of proteomic databases are available which allow mass spectrometry based experimental data upload, and subsequent comparison to other datasets contained in the database, such as PRIDE (http://www.ebi.ac.uk/pride/)[16]. Several other MS data repositories (namely Tranche and PeptideAtlas) have combined with PRIDE to form the Proteome Xchange (http://www.proteomeexchange.org) which enables submission from a single webpage and the combination of the data from all three repositories. In depth analytics on this data has not been performed- comparisons are mainly based around protein identification, and classification.

Quantitative comparisons of datasets in this forum aren't possible but grouping/result set selection according to numerous meta data and protein identifiers is possible. Absolute quantification comparisons with experimental datasets is possible through PaxDB (http://pax-db.org)[17] which not only contains data for most model organisms but has correlated absolute quantitation information from 28 datasets, and computed the average parts per million value for thousands of proteins. These data can be searched 100 identifiers at a time.

Using the MaxQuant software for data processing allows the grouping and separation of data from individual MS analyses [14, 15]. MaxQuant can combine data from all the protein fractions from a sample (if it has been pre-fractionated before MS), and can separate different samples from different conditions, but combine and output the results in one excel sheet. This facilitates direct comparison between all samples with all ratio/intensity data present.

When the appropriate population statistical analyses have been performed and a statistically valid significance cut-off has been calculated, the candidates for up- or down-regulated proteins from each group can be identified. When performing analysis of proteome dynamics, these results can also be compared with other variables. For example, a cell fractionation experiment performed, in conjunction with a time course of a drug treatment. Time course data can also be analysed to determine trends. It is important to have a zero time point, to describe the basal protein level, and use this to normalise values from the later time points followed by detection and grouping of trends. Most proteins will show little or no change over time but specific groups may show trends, for example reflecting regulation as a result of cell cycle, which appear as one or more peaks/troughs (figure 8) that can be identified by clustering analysis (this analysis was done with StatistiXL (http://www.statistixl.com/features/cluster.aspx) and further correlated with other data, such as GO terms or protein network information (Network analysis was done with String data base analysis http://string-db.org/[18]) . In the example shown, network analysis of the proteins found to have similar expression trends indicated that that the proteins identified were linkers between 2 or more functional networks, showing the transfer of effect through regulation, over time. With any other kind of grouping, such as for example Go term or subcellular location, this association between known networks would not be determined; it is only seen in the regulation trend association.

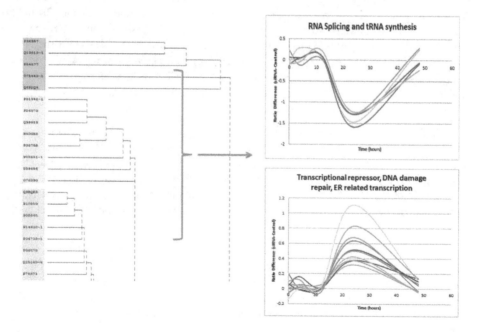

Figure 7. Hierarchical clustering of protein ratios over time, leading to effective grouping of expression trends. This kind of trend grouping and analysis was not possible by grouping according to GO terms or cellular location, or network association. Network analysis of these clustered groups **after** hierarchical clustering is advisable however, as interaction between known networks is often identified.

7. Conclusions

Life scientists working in the proteomics field have had the privilege of being at the cutting edge of an emerging technology that has opened up new possibilities for improving experimental design and data analysis. As proteomics can be "characterized more by its diversity than a common methodological or subject orientation"[1] the applications developed to accommodate this diversity should be made available and accessible to the wider scientific community.

The methods described here allow the description and measurement of protein-protein interactions, changes in proteome localisation and rates of synthesis and degradation. While the bench-top methodologies are relatively straightforward, the key to harnessing the biological value of the experiments often lies in the methods used to analyse the resulting data. We recommend systematic recording and management of all data, from all experiments. Systematic recording of detailed meta data can be used to extract information and obtain new results through a comparison of data trends across many different and often unrelated experiments. We term this approach, 'Super Experiments'.

All protocols discussed above can be found on greproteomics.lifesci.dundee.ac.uk and www.lamondlab.com websites.

8. Supplementary methods

8.1. SILAC-Stable Isotope Labelling of Amino acids in Culture

The following protocol provides a step by step guideline for preparing SILAC media and growing labelled cells in tissue culture.

Media can be bought ready or be made by the user, prior to use.

For media:

1. DMEM or RPMI minus arginine, lysine and methionine.
   Order no: contact your local sales rep
2. Dialyzed FBS (fetal calf serum)
   Order no: Invitrogen, cat no S181D (500ml)
3. Standard amino acids (ROKO media for control)
   Order no: Sigma, L-Arginine (A8094, 25g), L-lysine (L8662, 25g), L-Methionoine (M5308, 25g)

N/B: It is advisable to prepare 500µl aliquots of amino acids in PBS which can be stored at -20°C. Add 500µl aliquot of each when preparing SILAC media.

Stock concentrations: Arg0: 84mg/ml
                          Lys0: 146mg/ml
                          Met0: 30mg/ml

R6K4 and R10K8 amino acids- please note all amino acids are purchased via Cambridge Isotope Lab (CIL; North America, www.isotope.com) for UK see http://www.cgkas.com

Amino Acid	Symbol	Cat. No	Pack Size
L-arginine-HCL (U-13C6, 98%)	R8	CLM-2265	0.5g
L-arginine-HCL (U-13C6, 98% : 15N4, 98%)	R10	CNLM-539	0.5g
L-lysine-2HCL (U-13C6, 98%)	K6	CLM-2247	0.5g
L-lysine-2HCL (U-13C6, 98% : 15N2, 98%)	K8	CNLM-291	0.5g

These will make enough SILAC media for approximately 12 bottles of media however you can buy smaller amounts of the amino acids if you only plan to do 1 or 2 experiments.

Amino Acid	Symbol	Cat. No	Pack Size
L-arginine-HCL (U-13C6, 98%)	R8	CLM-2265	0.1g
L-arginine-HCL (U-13C6, 98% : 15N4, 98%)	R10	CNLM-539	0.1g
L-lysine-2HCL (U-13C6, 98%)	K6	CLM-2247	0.1g
L-lysine-2HCL (U-13C6, 98% : 15N2, 98%)	K8	CNLM-291	0.1g

4. Cell Dissociation Buffer
    Order no: Invitrogen, cat no. 13151-014 (100ml)

When passaging cells it is very important to NOT USE trypsin!! As this may provide a pool of unlabelled amino acids)

Preparing the SILAC media (500ml):

To 500ml DMEM/RPMI media add:

1. 500ml DMEM/RPMI media
2. 50ml dialysed FBS
3. 5.5ml Pen/Strep (and/or other antibiotics, if desired)
4. 0.5ml Met0 stock
5. 0.5ml Arg stock (R0, R6, R10)
6. 0.5ml Lys stock (K0, K4, K8)

Mix well then filter through 0.22µm sterile filter. Store at 4°C.

Cells should be grown for a minimum 6 passages for complete labelling.

8.2. Cellular fractionation protocol

This protocol will provide an effective technique to fractionate a variety of different cell types into cytoplasmic, nucleoplasmic and nucleoli fractions. The exact recipes for the solutions required throughout the protocol are provided at the end.

We have also included a shortened version of the protocol that will give only cytoplasmic and nuclei fractions.

N/B: Normal fractionation requires 5-15 x14cm circular dishes of completely confluent cells.

Cytoplasmic, Nucleoplasmic and Nucleoli fractionation

1. From confluent dishes. Trypsinise cells and spin in centrifuge for 4mins at 1000rpm. Wash pellet with PBS and spin again.
2. Re-suspend pellet in 5ml of ice-cold Buffer A (see Buffer A recipe). Incubate cells on ice for 5mins.
3. Transfer re-suspended pellet into a pre-chilled 7ml dounce homogeniser and break cells open using 10 strokes of a tight pestle.
4. Centrifuge dounced cells for 5mins at 4°C, 1000rpm. Retain supernatant as cytoplasmic fraction.
5. Re-suspend pellet in 3ml of S1 (0.25M Sucrose, 10mM $MgCl_2$) and layer over a 3ml cushion of S2 (0.35M Sucrose, 0.5mM $MgCl_2$) by slowly pipetting S1 solution on top of S2.
6. Centrifuge for 5mins at 4°C, 2500rpm.
7. Remove supernatant (retain if necessary) and re-suspend in 3ml of S2 (0.35M Sucrose, 0.5mM $MgCl_2$) and sonicate for 6 x 10 secs (with a 10 sec rest on ice between each sonication) using a probe sonicator. (N/B: if a probe sonicator is not available a bath sonicator can be used providing samples are sonicated in an ice bath to prevent overheating)
8. Layer the sonicated sample over 3ml S3 (0.88M Sucrose, 0.5mM $MgCl_2$) again by pipetting solution slowly on top S3 layer. Spin samples for 10mins at 4°C, 3500rpm.
9. Retain supernatant as nucleoplasmic fraction!
10. Wash pellet by re-suspending in 500µl of S2 (0.35M Sucrose, 0.5mM $MgCl_2$) and spin for 5mins at 4°C, 3500rpm- this is the nucleoli fraction.

Nucleoli pellet can be stored in any volume of buffer at -80°C and can be spun out again using the same centrifugation parameters as step 10.

Cytoplasm and Nuclei fractions only

1. From confluent dishes. Trypsinise cells and spin in centrifuge for 4mins at 1000rpm. Wash pellet with PBS and spin again.
2. From confluent dishes. Trypsinise cells and spin in centrifuge for 4mins at 1000rpm. Wash pellet with PBS and spin again.
3. Transfer re-suspended pellet into a pre-chilled 7ml dounce homogeniser and break cells open using 10 strokes of a tight pestle. Centrifuge dounced cells for 5mins at 4°C, 1000rpm. Retain supernatant as cytoplasmic fraction.
4. Re-suspend nuclear pellet in 3ml of S1 (0.25M Sucrose, 10mM $MgCl_2$) and layer over a 3ml cushion of S2 (0.35M Sucrose, 0.5mM $MgCl_2$) by slowly pipetting S1 solution on top of S2.
5. Centrifuge for 10mins at 4°C, 3500rpm and retain pellet as nuclear fraction.

Nuclear pellet can be stored in any volume of buffer at -80⁰C and can be spun out again using the same centrifugation parameters as step 5.

8.2.1. Solutions

Stock	mM (final)
1M HEPES, Ph 7.9	10
1M MgCl₂	1.5
2.5M KCl	10
1M DTT	0.5
dH₂O	Up to 10ml
Protease inhibitor	1 mini EDTA-free COMPLETE tablet

Table 1. Buffer A (10ml stock) is a hypotonic buffer that causes the cells to swell to they can be effectively broken open by dounce homogenizing.

Stock	mM (final)
2.5M Sucrose	0.25
1M MgCl₂	10
dH₂O	Up to 20ml
Protease inhibitor	1 mini EDTA-free COMPLETE tablet

Table 2. S1 (0.25M Sucrose, 10mM MgCl₂) 20ml

Stock	mM (final)
2.5M Sucrose	0.35
1M MgCl₂	0.5
dH₂O	Up to 40ml
Protease inhibitor	2 mini EDTA-free COMPLETE tablets

Table 3. S2 (0.35M Sucrose, 0.5mM MgCl₂), 40ml

Stock	mM (final)
2.5M Sucrose	0.88
1M MgCl₂	0.5
dH₂O	Up to 20ml
Protease inhibitor	1 mini EDTA-free COMPLETE tablet

Table 4. S3 (0.88M Sucrose, 0.5mM MgCl₂), 20ml

Up-scale volumes as necessary.

Stock	mM(final)
1M Tris, pH7.5	50
5M NaCl	150
10% NP-40	1%
10% Deoxycholate	0.5%
Protease inhibitor	1 mini EDTA-free COMPLETE tablet

Table 5. RIPA buffer- used frequently to prepare cellular lysate (10ml)

8.3. Immuno-precipitation protocol

This technique is very useful in the purification of a protein of interest. The technique works through the formation of an antigen: antibody complex which is attached to agarose/sepharose/metallic bead. The bead coupled to an antibody provide a matrix to which the protein of interest can bind allowing the other undesired components of the whole cell extract to be washed away. The eluted sample from the beads can then be further processed by gel electrophoresis and MS.

The protocol that follows is a very generic standard procedure presented as an initial recommendation for those who have not performed or optimised an IP previously.

Reagents required.

IP buffer:
20mM Tris-HCl pH 7.5
150mM NaCl
1mM EDTA
0.05% Triton X-100
5% glycerol
Protease Inhibitor cocktail tablets (Roche, cat. 11-873-580-001). 1 per 50ml buffer.

Glycine elution buffer:
100mM Glycine pH2.5 (adjusted with HCl)

Standard elution buffer: LDS sample buffer (invitrogen, cat. NP0007. Diluted 4x buffer 1:1 with milliQ to obtain 2x solution.

8.3.1. Method

N.B: All bead spin downs are done at 2000rpm for 2mins at 4°C.

1. Place whole cell extract aliquot in a round-bottomed vial to ensure good mixing. Add antibody to the required specific dilution for what you're using (you may need to consult your information booklet for antibody dilution guidelines.) When using cell

fractions use 200µl of Cytoplasmic protein solution, and 50µl of Nucleoplasmic protein solution (up-scaling as required).

2. Incubate between 0.5hours and overnight at 4°C, rotating.

NB. Perform all of the following steps on ice. Keep IP Buffer on ice also.

3. Wash beads with 1ml of IP buffer and spin down. Repeat. Re-suspend the beads in a 1:1 ratio with IP buffer. (i.e if 25µl of beads then 25µl of IP buffer)
4. Add 50µl of bead slurry to each Ab-lysate sample and rotate for 1-3 hours at 4°C.
5. Spin down beads. Retain the supernatant as this contains the unbound proteins.
6. Wash beads 3x with 1ml IP buffer. Vortex for 1 min before spinning down the beads.
7. Completely remove all liquid from the beads using gel loading tips then elute the bound proteins with either 2x 30µl aliquots of 2x LDS sample buffer (shaking for 5mins, at 95°C each time, for running samples on gels) or 2x 30µl glycine buffer (shaking for 10mins, at room temp each time, for doing in-solution digest).

N.B: if glycine buffer is used then it will result in a sample with an acidic pH. This needs to be neutralised so that further analysis can be done. Neutralisation of the sample can be done by slow, drop-by-drop addition of 1M Tris.HCl, pH 7.5. pH strips or LDS buffer (acidic pH will cause LDS buffer to turn yellow) colour can be used to check pH. In the case of in-solution digest the protein will need to be precipitated- in which case pH adjustment is not required.

8. Run both unbound and bound protein samples on a 1D 4-12% BisTris gel to provide a complete comparison. In Gel Digestion protocol can then be undertaken.

For further details on IPs and analysis with Mass Spec see the following;

*   Boulon, S., Ahmad, Y., Trinkle-Mulcahy, L., Verheggen, C., *et al.*, Establishment of a Protein Frequency Library and Its Application in the Reliable Identification of Specific Protein Interaction Partners. *Molecular & Cellular Proteomics* 2010, *9*, 861-879.
*   Trinkle-Mulcahy, L., Boulon, S., Lam, Y. W., Urcia, R., *et al.*, Identifying specific protein interaction partners using quantitative mass spectrometry and bead proteomes. *The Journal of Cell Biology* 2008, *183*, 223-239.
*   Ten Have S, Boulon S, Ahmad Y, Lamond AI. Mass spectrometry-based immuno-precipitation proteomics - The user's guide. Proteomics. 2011 Mar;11(6):1153-9. doi: 10.1002/pmic.201000548. Epub 2011 Feb 16.

Author details

Sara ten Have, Kelly Hodge and Angus I. Lamond
The Centre for Gene Regulation and Expression, College of Life Sciences, University of Dundee, Dundee, Scotland, UK

9. References

[1] Lamond, A. I., Uhlen, M., Horning, S., Makarov, A., *et al.*, Advancing cell biology through proteomics in space and time (PROSPECTS). *Mol Cell Proteomics* 2012, *11*, O112 017731.

[2] Ahmad, Y., Boisvert, F. M., Lundberg, E., Uhlen, M., Lamond, A. I., Systematic analysis of protein pools, isoforms, and modifications affecting turnover and subcellular localization. *Mol Cell Proteomics* 2012, *11*, M111 013680.

[3] Boisvert, F. M., Ahmad, Y., Gierlinski, M., Charriere, F., *et al.*, A quantitative spatial proteomics analysis of proteome turnover in human cells. *Mol Cell Proteomics* 2012, *11*, M111 011429.

[4] Boulon, S., Ahmad, Y., Trinkle-Mulcahy, L., Verheggen, C., *et al.*, Establishment of a protein frequency library and its application in the reliable identification of specific protein interaction partners. *Mol Cell Proteomics* 2010, *9*, 861-879.

[5] Boisvert, F. M., Lamond, A. I., p53-Dependent subcellular proteome localization following DNA damage. *PROTEOMICS* 2010, *10*, 4087-4097.

[6] Larance, M., Kirkwood, K. J., Xirodimas, D. P., Lundberg, E., *et al.*, Characterization of MRFAP1 turnover and interactions downstream of the NEDD8 pathway. *Mol Cell Proteomics* 2012, *11*, M111 014407.

[7] Deeb, S. J., D'Souza, R., Cox, J., Schmidt-Supprian, M., Mann, M., Super-SILAC allows classification of diffuse large B-cell lymphoma subtypes by their protein expression profiles. *Mol Cell Proteomics* 2012.

[8] Venter, J. C., Adams, M. D., Myers, E. W., Li, P. W., *et al.*, The Sequence of the Human Genome. *Science* 2001, *291*, 1304-1351.

[9] Ong, S. E., Blagoev, B., Kratchmarova, I., Kristensen, D. B., *et al.*, Stable isotope labeling by amino acids in cell culture, SILAC, as a simple and accurate approach to expression proteomics. *Mol Cell Proteomics* 2002, *1*, 376-386.

[10] ten Have, S., Boulon, S., Ahmad, Y., Lamond, A. I., Mass spectrometry-based immuno-precipitation proteomics - the user's guide. *PROTEOMICS* 2011, *11*, 1153-1159.

[11] Boisvert, F.-M., Lam, Y. W., Lamont, D., Lamond, A. I., A Quantitative Proteomics Analysis of Subcellular Proteome Localization and Changes Induced by DNA Damage. *Molecular & Cellular Proteomics* 2010, *9*, 457-470.

[12] Schwanhausser, B., Gossen, M., Dittmar, G., Selbach, M., Global analysis of cellular protein translation by pulsed SILAC. *PROTEOMICS* 2009, *9*, 205-209.

[13] Boisvert, F.-M., Ahmad, Y., Gierliński, M., Charrière, F., *et al.*, A quantitative spatial proteomics analysis of proteome turnover in human cells. *Molecular & Cellular Proteomics* 2011.

[14] Cox, J., Mann, M., MaxQuant enables high peptide identification rates, individualized p.p.b.-range mass accuracies and proteome-wide protein quantification. *Nat Biotechnol* 2008, *26*, 1367-1372.

[15] Schaab, C., Geiger, T., Stoehr, G., Cox, J., Mann, M., Analysis of high accuracy, quantitative proteomics data in the MaxQB database. *Mol Cell Proteomics* 2012, *11*, M111 014068.

[16] Vizcaíno, J. A., Côté, R., Reisinger, F., Barsnes, H., *et al.*, The Proteomics Identifications database: 2010 update. *Nucleic Acids Research* 2010, *38*, D736-D742.

[17] Wang, M., Weiss, M., Simonovic, M., Haertinger, G., *et al.*, PaxDb, a database of protein abundance averages across all three domains of life. *Molecular & Cellular Proteomics* 2012.

[18] Szklarczyk, D., Franceschini, A., Kuhn, M., Simonovic, M., *et al.*, The STRING database in 2011: functional interaction networks of proteins, globally integrated and scored. *Nucleic Acids Res* 2011, *39*, D561-568.

Permissions

The contributors of this book come from diverse backgrounds, making this book a truly international effort. This book will bring forth new frontiers with its revolutionizing research information and detailed analysis of the nascent developments around the world.

We would like to thank Dr. Germana Meroni and Dr. Francesca Petrera, for lending their expertise to make the book truly unique. They have played a crucial role in the development of this book. Without their invaluable contribution this book wouldn't have been possible. They have made vital efforts to compile up to date information on the varied aspects of this subject to make this book a valuable addition to the collection of many professionals and students.

This book was conceptualized with the vision of imparting up-to-date information and advanced data in this field. To ensure the same, a matchless editorial board was set up. Every individual on the board went through rigorous rounds of assessment to prove their worth. After which they invested a large part of their time researching and compiling the most relevant data for our readers. Conferences and sessions were held from time to time between the editorial board and the contributing authors to present the data in the most comprehensible form. The editorial team has worked tirelessly to provide valuable and valid information to help people across the globe.

Every chapter published in this book has been scrutinized by our experts. Their significance has been extensively debated. The topics covered herein carry significant findings which will fuel the growth of the discipline. They may even be implemented as practical applications or may be referred to as a beginning point for another development. Chapters in this book were first published by InTech; hereby published with permission under the Creative Commons Attribution License or equivalent.

The editorial board has been involved in producing this book since its inception. They have spent rigorous hours researching and exploring the diverse topics which have resulted in the successful publishing of this book. They have passed on their knowledge of decades through this book. To expedite this challenging task, the publisher supported the team at every step. A small team of assistant editors was also appointed to further simplify the editing procedure and attain best results for the readers.

Our editorial team has been hand-picked from every corner of the world. Their multi-ethnicity adds dynamic inputs to the discussions which result in innovative outcomes. These outcomes are then further discussed with the researchers and contributors who give their valuable feedback and opinion regarding the same. The feedback is then collaborated with the researches and they are edited in a comprehensive manner to aid the understanding of the subject.

Apart from the editorial board, the designing team has also invested a significant amount of their time in understanding the subject and creating the most relevant covers. They scrutinized every image to scout for the most suitable representation of the subject and create an appropriate cover for the book.

The publishing team has been involved in this book since its early stages. They were actively engaged in every process, be it collecting the data, connecting with the contributors or procuring relevant information. The team has been an ardent support to the editorial, designing and production team. Their endless efforts to recruit the best for this project, has resulted in the accomplishment of this book. They are a veteran in the field of academics and their pool of knowledge is as vast as their experience in printing. Their expertise and guidance has proved useful at every step. Their uncompromising quality standards have made this book an exceptional effort. Their encouragement from time to time has been an inspiration for everyone.

The publisher and the editorial board hope that this book will prove to be a valuable piece of knowledge for researchers, students, practitioners and scholars across the globe.

List of Contributors

Gabriela N. Tenea and Liliana Burlibasa
Department of Genetics, University of Bucharest, Romania

Fadhl M. Al-Akwaa
Biomedical Eng. Dept., Univ. of Science & Technology, Sana'a, Yemen

Bregje Wertheim
Evolutionary Genetics, Centre for Ecological and Evolutionary Studies, University of Groningen, Groningen, The Netherlands

Peter Ricke and Thorsten Mascher
Ludwig-Maximilians-University Munich, Germany

Hua Bai
Department of Ecology and Evolutionary Biology, Brown University, USA

Bushra Tabassum, Idrees Ahmad Nasir, Usman Aslam and Tayyab Husnain
National Centre of Excellence in Molecular Biology (CEMB), University of the Punjab, Lahore, Pakistan

Deepali Pathak and Sher Ali
National Institute of Immunology, Aruna Asaf Ali Marg, New Delhi, India

Francesco Panara and Ornella Calderini
CNR (National Council of Research) – Institute of Plant Genetics, Perugia, Italy

Andrea Porceddu
University of Sassari, Italy

Sara ten Have, Kelly Hodge and Angus I. Lamond
The Centre for Gene Regulation and Expression, College of Life Sciences, University of Dundee, Dundee, Scotland, UK

Printed in the USA
CPSIA information can be obtained
at www.ICGtesting.com
JSHW011404221024
72173JS00003B/415